W9-CZN-536

The Science *of* Ocean Waves

The Science *of* Ocean Waves

Ripples, Tsunamis, and Stormy Seas

J. B. Zirker

The Johns Hopkins University Press
Baltimore

© 2013 Johns Hopkins University Press
All rights reserved. Published 2013
Printed in the United States of America on acid-free paper
9 8 7 6 5 4 3 2 1

Johns Hopkins University Press
2715 North Charles Street
Baltimore, Maryland 21218-4363
www.press.jhu.edu

Library of Congress Cataloging-in-Publication Data

Zirker, Jack B.
    The science of ocean waves : ripples, tsunamis, and stormy
seas / J. B. Zirker.
        pages ; cm
    Includes index.
    ISBN 978-1-4214-1078-4 (hardcover : alk. paper) — ISBN
978-1-4214-1079-1 (electronic) — ISBN 1-4214-1078-8 (hard-
cover : alk. paper) — ISBN 1-4214-1079-6 (electronic)
    1. Ocean waves.   2. Tsunamis.   I. Title.
    GC211.2.Z57 2013
    551.46'3—dc23
                                            2012051092

A catalog record for this book is available from the British
Library.

*Special discounts are available for bulk purchases of this book. For
more information, please contact Special Sales at 410-516-6936 or
specialsales@press.jhu.edu.*

Johns Hopkins University Press uses environmentally friendly
book materials, including recycled text paper that is composed
of at least 30 percent post-consumer waste, whenever possible.

To Frances Cleveland,
whose critical judgment
helped to make this a better book

Contents

Preface

Some of my best memories of Hawaii are of watching the surfers at the Banzai Pipeline, on the north shore of Oahu. In the months between November and January, waves 10 meters (m) high or more roll in majestically, curl, and break with awesome power. These waves draw a dedicated band of top-flight surfers, who come to compete or just to test their skills.

But come back during a gale and see the power of the ocean when it is fully aroused by strong winds. Then the surf is really spectacular, with breakers that crash with a sonic boom and flood up the beach, carrying everything before them. In a hurricane, it is not worth your life to remain too near the shore.

Powerful ocean waves fascinate the public, and they have made a lot of news lately. We all remember the terrible loss of life and property that Hurricane Katrina caused in 2005. Much of the damage on the Gulf Coast was caused by battering waves that rode up a storm surge to a height of 9 m.

Then there was the tsunami launched by the great Sumatran earthquake in December 2004. At Aceh, near the epicenter, a wave of 30 m (98 ft) crashed onshore and obliterated the town. This impulsive wave crossed the Indian Ocean and killed over 200,000 people in 14 countries.

But the great tsunami that crushed the shore of Japan in March 2011 and inundated the Fukushima nuclear power plant was in some ways the scariest of recent events. The combination of a magnitude 9.0 earthquake, a 10-m tsunami, and the prospect of a core meltdown was a scenario usually seen only in science fiction.

Perhaps the most awesome waves are the so-called rogues or freaks that can rise up out of a moderate sea to heights of 20 m or more. In 1942, for example, the giant passenger ship *Queen Mary* was carrying 16,000 troops to

Surfing at the north shore of Oahu. (Photo 13438619, dreamstime.com.)

England. The ship was hit by a 28-m-high rogue wave that rolled the huge liner to an angle of 52 degrees. A few degrees more might have capsized the vessel. Such freak waves were thought to be extremely rare events, but radar-equipped satellites have since disproved that comfortable assumption.

Most ships lost at sea are wrecked by "ordinary" storm waves, however. The North Atlantic in winter is notorious for 10-m seas that persist for days. In the Drake Passage, between South America and Antarctica, waves commonly reach heights of 10 m and more, bedeviling the ships that try to round Cape Horn.

Powerful waves like these pose a real threat to shipping, and the maritime nations of the world have organized to cope with them. First and foremost they have sponsored research programs aimed at improving wave forecasting methods. Several forecasting centers now produce hourly or daily maps of wave heights to guide mariners at sea. In addition, satellite radars are deployed to monitor storm conditions.

In this book we'll look at all sorts of topics having to do with waves. I begin by introducing the properties of waves (without equations) and the physics that control them (chapters 1–4). Along the way we'll learn how a

Rogue wave hitting oil tanker *Overseas Chicago*, headed south from Valdez, Alaska, 1993. The ship was running in about 25-foot seas when the 60-foot wave struck it broadside on the starboard side. Photo by Captain Roger Wilson. (Courtesy of National Oceanic and Atmospheric Administration/Department of Commerce)

blustery wind generates ocean waves, how storm waves propagate, and how weak waves differ from stronger waves.

In Chapter 5, I describe some of the massive experiments oceanographers have carried out at sea and in laboratories to test their theories. Chapter 6 recalls the progress oceanographers have made in forecasting wave heights and directions. We'll see how radar works and how satellites are used to monitor great storms.

In the second half of the book, I discuss the beauty and power of breaking waves, the origins of those unpredictable rogue waves, the devastating tsunamis, and the ocean-wide El Niño phenomenon. The ocean tides are less dramatic than storms, but they are essential to the maritime industry. We'll recall how tides are generated and how daily forecasts are made (chapter 11). I discuss the amazing symmetry of ship wakes, and we'll learn how the hull of a racing yacht is designed to reduce the resistance of the waves they produce (chapter 13). Finally in chapter 14, we'll look into the development of clever machines that could capture the energy of ocean waves and tides and produce electricity on an industrial scale.

We'll go beyond mere anecdotes and try to understand as much as possible about wave physics without using mathematics. That means we'll have to review some basic properties of waves and the way winds push waves to great heights. We'll begin with the simple stuff and build from there. Some topics are more difficult than others, so take your time reading these parts.

It will be an interesting trip, so hop aboard!

The Science *of* Ocean Waves

CHAPTER ONE

A Walk along the Beach

We can learn a lot about ocean waves just by looking. So before we become immersed in the intricacies of waves, let's just stroll along the shore and comment on what we see. It's a nice, sunny day, without much wind: a perfect day for the beach.

As we look out to sea, we see a long train of parallel, equally spaced waves approaching the shore, as is shown in figure 1.1. These waves were probably generated by the winds of the storm that passed far offshore a couple of days ago. The sea is still recovering from the storm.

But what exactly are we looking at? The sea is not pouring steadily up the beach like a broad river. If it were, we'd be drowned. Instead, as each wave collapses on the beach, the water sloshes back into the sea. So we realize that these waves are part of a *moving pattern* of humps and hollows that glides over the surface of the sea. This regular pattern is called a *swell*. Swell waves are usually low, only about a meter high or so, and have rounded tops. All the crests we see are nearly parallel to the shore, have about the same height, and extend sideways at least six or seven times the distance between crests.

I'd guess that in this swell, the distance between the crest and the trough (which is called the "height" of the wave) is about 1 m. We could estimate the distance between wave crests (which is appropriately called the wavelength) as about 10 m, or 33 feet. And if we timed the interval between crests as they pass that buoy out there, we'd find the "period": about 5 seconds for these waves. Divide one number by the other and we get the speed of the wave, about 2 m/s, or 7 km/h, or about 4 mph—the pace of a fast walk.

Fig. 1.1 Snapshot of a swell. (Photo 14568824, dreamstime.com.)

The Water under the Wave

There's a swimmer out there, floating on her back. Notice how she rises and falls rhythmically as each wave passes her. Although the waves are moving toward shore, she hardly advances shoreward. Her motion follows that of the water beneath her. It may seem surprising, but the water in a wave doesn't actually travel with the wave toward the shore; it just bobs up and down, practically in place. We'll talk more about this oscillating motion, and lack of forward motion, later on.

If this swimmer were to dive below the surface, she'd discover that the oscillations of the water gradually become weaker and weaker the deeper she dives. A few meters below the surface she would float in practically still water. Submariners are familiar with this phenomenon; they can escape a violent storm at the surface by diving deep enough to reach calm water.

Surf

Back on our beach, we see some kids playing in the surf zone where the small waves finally break. One little guy ventures out too far and gets knocked over by a wave. He's all right; he picks himself up and runs back up the beach. His little accident reminds us, however, that a breaking wave carries a punch. Or in more technical terms, a wave carries the energy the wind gave it and releases that energy when it breaks. When a wave breaks, its energy accelerates the water, which then has enough momentum to knock you over. If you've ever waded out through the surf to reach quiet water beyond the breaking waves, you'll understand what I mean.

This beach that we're walking along is curved in a deep arc, a C shape maybe 2 km long. As we walk toward the rocky point at the far end, we keep a sharp eye on the waves offshore. We notice that everywhere along the beach, the waves come rolling in parallel to shoreline. Somehow the waves rolling in from the horizon *turn* so as to face the shore at every point. How is this possible? This effect is called *refraction*, and we'll learn how it works later on. Every type of wave (such as sound, seismic, or electromagnetic) exhibits refraction.

As we walk along, we notice that the appearance of the breaking waves changes from place to place. Where we started out, the beach sloped very gently into the water and the waves broke very gently. These were "spilling" waves; you can see an example in figure 1.2 (top).

Fig. 1.2 Top, a spilling wave; *bottom*, a plunging wave. (Photos 12068479, 3918008, dreamstime.com.)

Further along, the beach becomes steeper, and the crest of each wave curls and plunges forward as it reaches the beach (the "plunging" waves in fig. 1.2, bottom). Avid surfers look for a beach with just the right amount of slope to create a good plunging breaker. Finally, we reach a part of the beach that slopes very steeply away from a cliff, and here the waves barely rise up before smashing against the cliff. These are called "surging" waves.

Later on we'll examine this connection between the shape of breakers and the slope of the beach in more detail.

Playing with Waves

We pass two little girls who are dropping pebbles into a circular pool of water they've dug in the sand. As a pebble falls in the water, it creates a circular ripple that spreads out and *reflects* from the edge of the pool toward the center, as can be seen in figure 1.3A. This event is a small version of a tsunami! The pebble represents the undersea earthquake that launches a group of waves across the water. The waves cross the ocean and reflect back from a

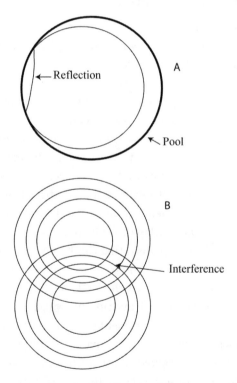

Fig. 1.3 A, circular waves arise when we drop a pebble into a pool. When they reach a border they are reflected. *B*, the cross-hatched interference pattern that arises after dropping two pebbles at the same time.

coast. This effect was seen in the Indonesian tsunami of 2004. It crossed the Pacific Ocean basin at 750 km/h and bounced off the east coast of Africa. Incidentally, *reflection* is another universal property of waves.

Before we leave the girls playing by the pool, watch what happens when they drop two pebbles at the same time. Now we have two circular patterns that expand outward and cross each other, as shown in figure 1.3B. When two crests overlap, the result is a taller crest; when a crest and a trough overlap, they cancel each other and the result is a draw. This *interference* of water waves is remarkable: they can pass over and through each other without disruption, but only if the waves have small heights compared with their wavelengths. Tall, steep waves can behave quite differently, as we shall see later on. Once again, interference is a behavior common to all types of waves.

Navigation by Wave Patterns

Let me digress from our stroll on the beach to note that interference patterns in the ocean have been used in a very practical application: navigation. The natives of Micronesia and Polynesia were famous for the long voyages they made in open canoes across hundreds and thousands of miles of empty ocean. They could be out of sight of land for many weeks, and yet they could locate a tiny island in the midst of the vast ocean.

To navigate they used a variety of aids, such as the stars, cloud formations, winds, currents, and the flight of birds. In addition, the natives of the Marshall Islands in the western Pacific developed a special skill. They learned to read the interference patterns of swells that were driven by the prevailing northeast trade winds. Swells bend around islands and spread out in the channels between them. The overlap of swells from different directions produces a distinctive interference pattern that can help to fix your location.

The Marshall Islanders preserved their knowledge of the sea in so-called stick charts, which were passed down through the generations. The charts were made of strips of coconut leaf midrib and wood. Small cowrie shells were attached to the framework to represent individual islands. Curved strips represented the zones where interference patterns could be found. Other strips represented currents. A skilled navigator would orient the chart with the sun or stars and look for a particular interference pattern to guide his voyage. A simple but effective scheme!

A Bird's Eye View

Now let's climb to the top of the high cliff that looks down on the shore. From there we can see how a swell interacts with itself and with a small island offshore. In figure 1.4 a swell is traveling from the lower right to the upper left. As it brushes against the mainland, the right ends of its wave turn slightly (*refract*) to face the cliff (notice the little bends in the ends). Then these refracted waves *reflect* off the promontory and *interfere* with the oncoming waves.

We can also see a good example of *diffraction* as the swell squeezes between the island and the mainland: a series of spreading circular arcs. Finally we see another swell entering from the left and interfering with the diffracted waves. Once again, reflection, refraction, diffraction, and interference are basic processes that all types of waves exhibit. So not only ocean waves, but also sound waves, light waves, and seismic waves show them.

Ah, but now the wind is picking up. We're about to see how the sea changes under a rising wind. At first we see small waves building on top of the existing swell. These ripples break up almost immediately into small whitecaps because of the force of the wind. This is what's called a *choppy* sea, or a *chop* for short.

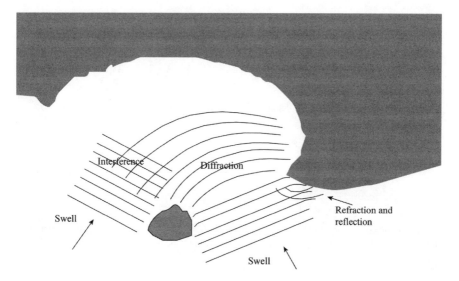

Fig. 1.4 Looking down on the sea from a cliff, we can see several phenomena common to all types of waves: reflection, refraction, interference, and diffraction.

Fig. 1.5 An example of a "sea," a jumble of short, high, pointed crests. (Photo 19376143, dreamstime.com.)

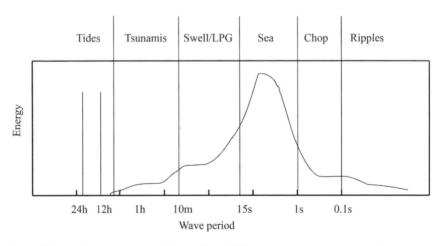

Fig. 1.6 Types of waves, arranged by period. (*LPG* refers to long-period gravity waves.) The curve indicates the amount of energy each type possesses.

Now the wind is rising very quickly; we are having a fierce squall. After a short while the sea is churned into chaos, with tall waves running in directions away from the wind and breaking into whitecaps. Sailors would call this a "sea" (fig. 1.5). Finally, in a gale or hurricane the ocean becomes a "fully developed sea." Now the wave crests have sharp pointed tops, are irregular in height, and extend sideways only a few wavelengths. Short waves are piled on top of long waves, and the sea surface is bouncing up and down erratically. A small boat could easily be swamped in such a sea.

This is a good place to summarize the basic properties of ripples, chop, seas, swells, tsunamis, tides, and other types of waves. In figure 1.6 we see these waves arranged in order of period. The curve indicates the amount of energy each type possesses in the sea.

Well, the wind has turned cold. We'll meet all these waves again, along with some scientists who have studied them, but for now it's time to move on.

CHAPTER TWO

What Exactly Is a Wave?

Everybody knows what a wave is. At least, we recognize one when we see it. When asked to imagine a wave, most people think of ocean waves, those majestic waves that roll steadily toward the shore or the chaotic waves in a stormy sea. Most people also know that other kinds of waves exist: light waves, sound waves, and earthquake (seismic) waves, for example. But when asked to describe what they see, most people are a little vague. Obviously something is moving but what exactly? Let's try a few thought experiments to answer the question.

Remember the game we used to play with dominoes as children? We'd stand them up on their short sides in a long row, taking care to space them apart by the same small distance. Then we'd touch the first in line and watch them all fall over, one after the other. We'd see a wave of some sort moving rapidly down the line. Very exciting! But what was moving? Each domino slumped on its neighbor and came to rest. On the other hand, the wave traveled very quickly to the end of the line.

Therefore we could say, in a general sense, that a wave is a traveling disturbance in some medium. The medium in this case was the line of dominoes; the disturbance was the tipping of each domino. That's not a bad definition. But we could also say that the wave was carrying a message from one domino to the next: "Lean away from me!" So in some sense it was the angle of leaning that was traveling down the line. Notice that no domino traveled with the wave down to the end of the line: each domino just tipped over and stopped. It is the message that moves, not the medium.

Strictly speaking, I would call this single domino wave a pulse, not a genuine wave. Genuine waves, in my opinion, involve some type of oscillation. A good example is provided by a child's toy, a Slinky.

The Slinky

A Slinky is simply a long, loosely coiled spring (see the image at the top of fig. 2.1). If we push and pull one end of the spring rhythmically in and out, we send waves of compression and rarefaction down the coil, a good analogy to pressure waves in a sound wave.

In figure 2.1 we see a time sequence of such a compression wave in the Slinky. Initially the Slinky is at rest, and each loop is in its rest position, equally spaced from its neighbors. Then at time T_1 we push in our end, creating a region of compressed loops. The compression moves down the spring. If we look carefully, we'll see why: each loop of the spring moves forward a small distance and pushes the next loop in line, and so the compression advances.

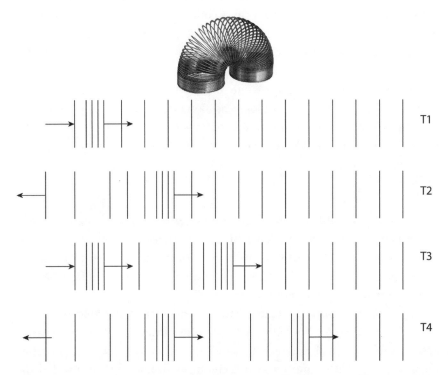

Fig. 2.1 A wave of compressions and rarefactions on a Slinky. At time T1 a compression is formed. It moves down the coil at time T2. At time T3, a rarefaction is formed and moves down the coil. These oscillations of loops about their rest position generate the wave we see.

In this forward motion, each loop overshoots its rest position: its momentum has carried it too far forward. But then the tension in the spring acts as a restoring force on the loop and causes it to rebound from its furthest advance. Once again the loop overshoots its rest position, and tension pulls it forward. Each loop therefore oscillates back and forth, along the direction of wave propagation.

A half period later (T_2) we pull back on our end of the spring and create a rarefaction, a region of low loop density. This time each loop *pulls* on its neighbor in the direction opposite to the direction of propagation. The rarefaction also propagates down the spring (T_3). Notice that although the loops are moving backward in their oscillation, the rarefaction is moving forward. The cycle repeats at time T_4. No loop travels from one end to the other. Only the wave energy travels that far.

The main point here is that the wave propagates because each element of the medium (the loops of the spring) communicates its oscillations to its neighbor downstream after a short delay. The speed of the wave, it turns out, is controlled by the stiffness of the spring: the harder it is to push or pull the loops, the faster is the wave.

Incidentally, the Slinky waves are *longitudinal* waves, meaning that the displacement of an element (a loop) was along the direction of propagation. The Slinky wave is a good model of a sound wave in air, a series of compressions and rarefactions. But we need a better model for water waves. A child's jump rope is a simple example.

A Jump Rope

Let's imagine two girls pulling gently on opposite ends of a long rope. Now let's watch as the girl at the left end, Louise, snaps her end up and down sharply, just once. We see a kink in the rope travel quickly along the rope to Rachel, at the other end (fig. 2.2A). When the kink arrives, Rachel's hand is snapped up and down, just once.

This kink was a *pulse*, a single isolated disturbance in the rope, not a true wave. But it served to show that that the pulse carried *energy* from Louise to Rachel, sufficient to shake Rachel's hand up and down. Her hand absorbed the energy, so that no pulse was reflected back to Louise. The speed of the pulse depends on how much tension the girls have put on the rope: the greater the tension, the faster the pulse. It also depends on how heavy the rope is: the heavier the weight, the slower the speed.

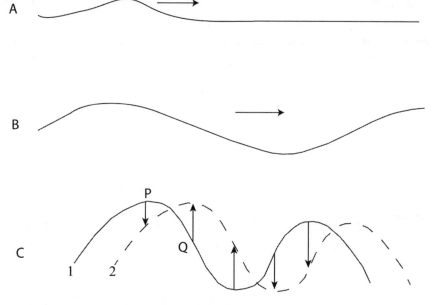

Fig. 2.2 Waves on a rope. *A*, a pulse; *B*, a true wave; *C*, a crest shifting rightward to the next part of the rope.

Next, we watch Louise shake her end of the rope rhythmically. She generates a true wave, a sequence of kinks that travel steadily to Rachel (fig. 2.2B). Rachel's hand is shaken as before; she is absorbing the energy that Louise is pumping into the rope. If Louise shakes the rope more frequently, the distance between the kinks (the wavelength) becomes shorter, but the kinks travel the same speed as before. That's because the speed depends only on the rope's tension and weight.

While Louise is shaking her end of the rope, we should look carefully at the motion of kinks in the middle of the rope. In figure 2.2C we see the rope at two instants, 1 and 2. The arrows indicate the directions of the motion. The pieces of rope are definitely not moving bodily toward Rachel. They are merely oscillating up and down a short distance from their rest positions. And yet their vertical motions help to produce the horizontally traveling wave that we seem to see. How does this happen?

First let's agree that what we interpret as a wave is an apparent movement of the high spots in the undulating rope. The rope is not streaming toward Rachel as a whole; only the locations of the high spots are streaming.

We notice that during its oscillation, point Q exactly repeats each of point P's vertical motions, but with a short time delay. For example, when P is at the top of its rise and moving down (see the arrows), Q is still rising toward its top (fig. 2.2c). That means that Q will reach its top a short time after P has. The location of the high spot in this part of the rope will have moved from P to Q. We would interpret this shift as a forward movement of a wave.

It is exactly this time delay in the motion of neighboring segments that generates the wave that we find so eye-catching. Each segment lags the segment behind it, in Louise's direction, by the same delay. Therefore, we see that the closer a segment is to Rachel, the later it reaches the top of its rise. Basically, the wave is *a horizontally moving pattern* that arises from the motion of *vertically oscillating* elements.

Incidentally, this is an example of a *transverse* wave, in which the displacements of the medium (the rope) are perpendicular to the direction of wave propagation. Light is also a transverse wave. So are the vibrations of a guitar string and an ocean wave.

We now have the tools we need to talk about water waves, so let's move on.

The Construction of a Water Wave

Watching gentle ocean waves roll onto a shore can be hypnotic. It's a very restful pastime that I've enjoyed occasionally. But after you've been watching for some time, you may begin to wonder why the waves are so regular (as in fig. 1.1). The crests are spaced apart by a constant distance (the wavelength), they move toward shore at a constant speed, and they arrive at a constant interval (the period). Moreover, each wave has very nearly the same shape. It may remind us of the sine waves we studied in high school, except that the crests seem a bit sharper and the troughs a bit broader.

How is this regular pattern maintained, especially with no wind? What is happening under the surface of the water? And how are the different characteristics of the waves related? We can get a clue to the mystery by watching that swimmer offshore, who is floating on her back. As a crest approaches her, she rises at first and moves forward a bit, then sinks, and finally moves backward, in what looks like a circular motion. She's not surfing, not being carried forward with the surface wave; she and the water under her are just oscillating in place.

Clearly these incoming waves have some connection with oscillations of the water. To learn what the connection is we need to look under the surface.

Fortunately, two clever brothers did this for us almost 200 years ago and reported what they saw.

Under the Surface

Wilhelm Eduard Weber (1804–91) and his older brother Ernst Heinrich Weber were the first scientists to investigate water waves experimentally in a lab. In 1825 Wilhelm was a 20-year-old graduate student in physics at the University of Halle in Saxony, Germany, and an avid believer in experimenting. Ernst was already a professor of physiology at the University of Leipzig, interested in how blood flows through arteries, and he decided to draw on Wilhelm's proven experimental skills.

Wilhelm was easily persuaded. He set up several glass tubes with different diameters, pumped various fluids through a tube under precise pressures, and measured the rates of flow. In a long series of trials he and Ernst determined the viscosity (stickiness) of such fluids as mercury, water, and brandy. These simple experiments were so successful that the brothers probably celebrated by drinking the brandy. Then, in the flush of victory, they decided to tackle a more difficult problem, the motion of water waves.

So they set up a rectangular wave tank, a narrow channel 2 m long and 2 cm wide, filled 0.5 m deep with water. They could launch a train of small waves by dropping precise amounts of water into the tank at regular intervals (the "period"). Then they could measure the separation of the crests (the wavelength) and the speed of the crests. They could also observe the shapes of the waves through the glass sides of the tank.

The first thing they noticed was that the profile of the wave train resembled the familiar sine wave one learns about in high school, but with some differences: the peaks were a little sharper and the troughs were broader. Ocean waves are like that too. Second, the longer the period of the wave, the longer was its wavelength and the faster its speed, although they were unable to determine precise relationships. Third, the height of a typical wave (the vertical distance from crest to trough) could be made larger or smaller without changing the period or wavelength.

Their most interesting results concerned the motions of the water under the surface. To make these more visible, the Webers added small particles of flour to the water. Figure 2.3A shows a reconstruction of what they could have observed through the glass sides of the tank, a snapshot of a wave traveling to the right.

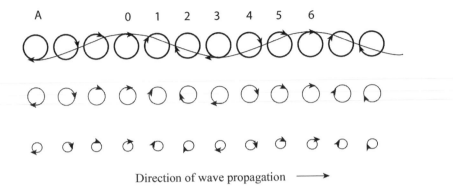

Direction of wave propagation ⟶

Fig. 2.3 Water blobs under the ocean's surface rotate in circles as a deep-water wave passes by. See text for description.

Under the surface, a traveling water wave looks like the inside of a fine clock, filled with carefully synchronized "gears." The gears are actually the vertical circular orbits of small blobs of water (let's just say the size of a blob is a small fraction of the wavelength). The orbits decrease in size the deeper one looks. Moreover, except for a very small drift in the forward direction, each orbit remains in its place as the crest passes by. (This is the reason the swimmer discussed earlier merely rocks up and down in the water without advancing toward the shore.)

In the illustration each blob revolves in its orbit at a constant angular speed in a clockwise direction and completes a revolution during one period of the wave. Moreover, all the blobs in a vertical column are at the tops of their orbits (and moving horizontally) at the same instant, just as a crest passes overhead. In fact, a crest is just the visible result of this coordinated rise of all the blobs. Similarly, when all the blobs in a vertical column are at their lowest points, they create a trough.

But there is more to this subtle motion. The blobs in each vertical column reach the tops of their orbits a little later than those in the column to their left (or upstream). So in figure 2.3A, the blobs in column 1 lag those in column 0 by a small amount—in this drawing, a sixth of a period. And the blobs in column 2 lag those in column 1 and so on down the line to the right.

When all the blobs in column 1 reach the tops of their orbits at the same moment, they stack up to form a new crest at the surface. You can see this in figure 2.3B. Here, we see the position of the wave (the dashed line) after a sixth of a period has elapsed. The blobs in column 1 are now at the tops of their orbits and have created a new crest. In effect, they have shifted the position of the crest to the right by one-sixth of a wavelength. The small arrows show the paths that blobs on the original wave have taken. And the process repeats: the crest continues to shift from one column to the next one downstream. In this way, the crests and troughs travel from left to right in the figure.

For convenience, I've drawn only seven orbits between two crests in figure 2.3A, but of course there is an orbit of the same size—an infinite number—for every point on the wave's profile. The orbits would overlap, but because of the way the blobs lag each other, they never collide. The motion of the fluid would be perfectly smooth. The same is true for the orbits below the surface of the water.

The Webers made an important discovery with their simple experiments. They revealed that there is a tight connection between the revolution of blobs of water in stationary orbits and the passage of a surface wave. Indeed, the wave and the orbits have the same period. The whole train of waves advances by one wavelength in one period of the oscillation. Physicists call this a traveling wave train. We can call it a swell.

The Weber brothers summarized their research in a massive 575-page monograph in which they compared their observations with the laws of fluid flow known at the time. It was the first major advance in the experimental study of waves. But as is often the case, brilliance in one area often carries over to a lifetime of brilliant works. In later life, Wilhelm collaborated with Carl Friedrich Gauss, a brilliant mathematician, in a comprehensive study of magnetism. They also invented the first telegraph system, but that is another story.

How Energy Propagates

You can watch a swell rolling onto the shore all day without any wind to push it. How is this possible? Sailors all know that the waves are created by a storm or a prevailing wind far offshore. The harder the wind blows and the greater the distance it acts on the water (the "fetch"), the higher the waves grow. But once the waves start rolling, there isn't much to slow them down because the internal friction of water (its viscosity) is very weak. So these wind-formed waves roll on until they crash on the beach. There they release the energy that the wind delivered to them.

We can think of a wave train as a conveyor belt that carries energy at a steady rate toward the shore. The basic elements in the conveyor belt are the orbiting blobs of water near the surface. Each blob stores and releases gravitational energy as it rises and falls in its orbit. A blob behaves like a child on a swing, storing gravitational energy as it swings up to its high point, and releasing it as it falls. Each falling blob delivers some of its stored energy to its neighbor downstream, causing it to rise, and so on in a chain reaction. A blob would come to rest after passing along its energy, except that it immediately receives another pulse of energy from its neighbor upstream.

In effect, the blobs are acting like the workers in a bucket brigade, passing energy down the line without moving far from their normal positions. So the net result is a train of waves that delivers a steady flow of energy to the beach. Because gravity is the key force that enables the water to store energy temporarily, ocean waves are called *surface gravity waves*.

There remains the mystery of how the rotating blobs of water got started, how they became synchronized, and why they don't propel waves away from the shore. In the next chapter we'll return to these questions. There we'll see how a prevailing wind imparted some of its energy to the water and also fixed the direction of rotation (clockwise or counterclockwise) of the circular orbits of the blobs. But first let's continue with these blobs.

Restoring Forces

At this point you should have a pretty good idea of how water blobs rotate within a traveling gravity wave. But I really haven't explained how gravity and pressure drive these motions. As we saw earlier, each blob moves both vertically and horizontally as it revolves in a circular orbit. Let's first consider the forces that control the vertical motions. There are two opposing forces in play: *gravity* (which always acts to pull a crest down and to deepen a trough) and water *pressure* (which builds up under the falling water and pushes against it). Gravity would cause a blob of water to fall indefinitely, but rising pressure from the surrounding water acts as the restoring force in the vertical oscillation of the water.

The motion resembles that of a child bouncing on a trampoline. Gravity pulls her down, and she hits the trampoline's fabric surface hard enough to stretch the coiled springs holding up the fabric. However, as the combined tension from the coiled springs increases, it eventually becomes larger than the force of gravity, causing the springs to contract sharply and launching her skyward again. In this analogy, the spring tension mimics the rising water pressure in a wave.

In figure 2.3B we can see how this works. We see two snapshots of a sinusoidal wave, taken a sixth of a period apart. The wave (the dashed curve) has moved a sixth of a wavelength to the right in this time. The arrows show the paths that some blobs took during this time. As the back of a crest collapses, the falling water increases the pressure underneath it. At position 1 the pressure increased enough to slow the collapse; at position 2 the excess pressure has reversed the fall and is raising a new crest to its maximum height. In this way the water oscillates vertically, with gravity pulling it down and the surrounding water pressure pushing it back up.

The *horizontal* oscillation of the water blobs is driven solely by oscillating horizontal pressures. These pressures are less obvious, but they account for the fact that the paths of the water blobs are circles and not merely vertical lines. At position 1 in figure 2.3B, a falling crest creates a horizontal pressure that is larger than the pressure at position 2. That difference of pressure pushes the blobs downstream a short distance and ensures the propagation of the crests.

The Vital Connections

Now let's get back to the question I posed earlier: what is the relationship between the wavelength and the period of a wave? For a quick and crude answer we can appeal to an analogy between pendulums and water waves because gravity governs both of their oscillations.

Galileo Galilei was the first to investigate pendulums in a laboratory. He was probably the most famous astronomer of his century but few people today realize that he was also a talented experimentalist. His reputation rests on his stunning astronomical discoveries, which include the moons of Jupiter, the rings of Saturn, and the changes in the shape of Venus's illuminated disk. But he also made important contributions to our understanding of how bodies move. In particular, he discovered the essential properties of pendulums in series of experiments.

According to a legend, he became fascinated with pendulums after noticing a chandelier in the Pisa cathedral swinging with a constant period. Later in his lab he learned that in order to double the period of a pendulum, you must make the length of the arm four times longer, and to triple the period you must make the arm nine times longer. This led ultimately to the basic equation that correlates the arm's length with the square of the period: thus, 2 squared is 4, and 3 squared is 9.

Now, to the extent that the analogy between gravity waves and pendulums is valid, we could identify the pendulum arm with the wave's wavelength and guess that the wavelength of a gravity wave increases as the square of the period. I'll admit that this explanation may seem like sleight of hand. For a sound physical explanation we'll need to turn to the long line of scientists and mathematicians who investigated water waves.

Sir Isaac Newton

Sir Isaac Newton was probably the first to propose a theory of gravity water waves. (Why isn't that surprising? He seems to have pioneered everything.) To Newton we are indebted for the law of gravitation and an explanation in terms of physical forces for the movement of the planets. He applied the same methods to water gravity waves.

We will meet his prolific contributions again in later chapters, since he made major contributions to the science of optics and independently invented differential calculus, which allowed future scientists to perform the mathematical calculations needed to truly understand wave dynamics. He was undoubtedly a genius. But he was also an alchemist who tried to convert lead into gold and was well-known in his time as a mystic and a secretive, eccentric person.

In his monumental treatise *Philosophiæ Naturalis Principia Mathematica* (1687) he published an approximate theory that yielded a prime result: the wavelength of a water wave increases as the square of its period. (Just as we guessed!) So, for example, a gravity wave with twice the period of another wave has a wavelength four times as long.

D'Alembert and Euler

Jean LeRond d'Alembert, a French mathematician born in 1707, shortly before Newton died, was the next in line to study waves. An illegitimate son of a wealthy man, whose mother abandoned him and whose father refused to acknowledge him publicly, he was pressured to become a priest by his father's family. Given this start, it would seem unlikely that he would ever have had any connection with the science of waves. But as he grew up, he strongly rejected theology ("rather unsubstantial fodder") and became a mathematician instead. He helped to build a mathematical theory of music, including the origin of overtones, the concepts of octaves and major and minor chords, and much more.

Eventually, in 1747, he became interested in how the strings on a violin make a specific tone when plucked. He chose a mental model used earlier by Johann Bernoulli (a Swiss mathematician from the illustrious Bernoulli family of scientists): a chain of beads connected by little springs. Like Bernoulli, he relied on Robert Hooke's law (Hooke was a British compatriot of Isaac Newton and a fellow natural philosopher), which says that the restoring force of a stretched spring is proportional to the amount of stretch. Then he imagined shrinking the size of each bead and increasing their number, while keeping the same total mass for the string. He also replaced the springs with a continuous tension in the string. In an important publication, he derived the wave equation that governs the motion of this system and found a general solution. It was the first mathematical analysis of string vibrations ever published.

D'Alembert demonstrated that a wave need not be periodic or have the shape of a sine wave or have such a thing as a wavelength. Indeed, a wave could consist of a single traveling hump of an arbitrary shape (like the pulse traveling in our domino example). He proved that the shape will remain unchanged as it progresses *only* if each point on it advances the same distance in a given time. With proper modifications, his wave equation would apply to any wave, water waves included. But somehow he didn't apply it to water. He went on to serve with Denis Diderot as co-editor of the *Encyclopédie*, a vast, sprawling work which aimed to summarize all of science and art—a far cry from his original destiny.

D'Alembert was followed by Leonhard Euler, a Swiss mathematician of the eighteenth century, who has a good claim to be the greatest mathematical physicist of his age. Euler made huge contributions to a vast array of topics in mathematics, including analytic geometry, calculus, trigonometry, and number theory. His advances in physics include many practical subjects, such as a 700-page study of the motion of the Moon; cannon ball ballistics; and the wave theory of light, water, and stringed instruments.

Perhaps Euler's most important work in physics was the derivation of the three equations that govern the motions of fluids that have zero friction (or viscosity). The equations state that in a fixed volume of fluid, mass is neither created nor destroyed, nor is energy created or destroyed; and the momentum of the mass is the result of the known external forces. The wave equation is a special case of these general equations of flow. Around 1760 Euler obtained solutions in terms of sine and cosine waves, but for some reason he didn't apply them to ocean waves.

I can't resist telling a little story about Euler. Despite his dedication to serious work, Euler loved puzzles. Just for fun he solved the problem of the Seven Bridges of Konigsburg. This city in Prussia sits astride the river Pregel. There are two islands in the river that are connected to the mainland by seven bridges. The citizens wanted to know whether it was possible to cross all seven bridges once and once only, in such a way as to end where one started.

Nobody had been able to prove, by actually walking the route, that it was possible. Euler decided to investigate the problem mathematically. He solved the puzzle in 1735, proving that such an optimum route was in fact impossible. And his solution sparked interest in a new subject in mathematics called topology. Similar problems arise in modern life, as when a traveling salesman wants the shortest air route among a fixed number of cities. But I digress.

Several outstanding French mathematicians of the eighteenth century pioneered in describing water waves, at least in special cases. Around 1786 Joseph-Louis Lagrange described long waves in shallow water and proved that the blobs of water move in flattened elliptical orbits and, indeed, become increasingly flattened toward the bottom of the water. (The orbits are circular in very deep water, as shown in fig. 2.3A.)

Then in 1816, the French Académie des Sciences Nationale offered a prize for the best mathematical work on the motion of waves on a deep ocean. Augustin-Louis Cauchy, only 25 years old, tired of his engineering work for Napoleon and ill from overwork, turned to pure mathematics. On sick leave and living at his parents' home in Paris, he tackled wave propagation and submitted his mathematical analysis of the circular surface waves that are launched by dropping a pebble in a pool. Siméon Denis Poisson, one of the judges, entered his own work, which seems like a conflict of interest. Cauchy won the Académie's Grand Prix anyway, despite the fact that his innovative mathematics was probably too abstruse for the judges to fully appreciate. However, his analysis methods, eventually improved by Poisson, have enabled future oceanographers to predict the evolution of any initial state of the water surface, as we will see in later chapters.

Gravity Waves: Airy's Theory

A comprehensive theory of "weak" water waves, whose heights are very much smaller than their wavelength, was achieved only in 1841. Sir George Biddell Airy, mathematician and Astronomer Royal of Great Britain for almost 50 years, worked out the behavior of weak gravity waves. His theory applies to freely running waves that are not being pushed by a wind, and it applies reasonably well to ocean swells.

Airy found that gravity waves behave differently in deep and in shallow water. Deep water, in this context, is at least half a wavelength deep. Airy easily confirmed Newton's result that in deep water, the wavelength increases in proportion to the square of its period. So, for example, waves with periods of 4 and 8 seconds have wavelengths of 25 and 100 m, respectively. Longer period waves have *much* longer wavelengths.

The speed of a deep water wave is proportional to its period. Therefore, long-period (or long-wavelength) waves travel faster than short-period waves. Waves with 4- and 16-second periods, for example, move at speeds of 6 and 25 m/s. The fact that the speed of a deep-water wave depends on its period (a property called *dispersion*) has important consequences. In a stormy sea, waves with many different periods are generated. Because of dispersion, the long-period waves race ahead of the shorter waves and arrive at the beach much earlier as a swell. Short-period waves also tend to die out much sooner.

Airy determined that waves in shallow water are a different breed altogether. By "shallow" he meant that the depth to the sea bottom is less than half the wavelength of the wave, typically near a shore. (With this definition, however, tsunamis, which have wavelengths of tens of kilometers, behave like shallow-water waves even in the deepest oceans!) A shallow-water wave moves in elliptical orbits under the surface, in contrast to the circular orbits of deep-water waves. In figure 2.3 I've assumed a deep-water wave.

Airy also showed that if the depth to the bottom is less than a twentieth of the wavelength, the speed of a wave depends only on the depth of the water. In this case, the wave's wavelength depends on both the period and the depth. So, for example in water 10 cm deep, waves with periods of 3 and 10 seconds would have the same speed, 0.99 m/s, and wavelengths of 3 and 10 m. And some tsunamis have traveled across the Pacific Ocean at 900 km/h!

The situation at a sloping beach is more complicated, and Airy's theory doesn't apply. On the run up to the beach, the wavelength and speed decrease, the height grows, and finally the wave tips over and crashes as surf. We'll discuss these breaking waves in a later chapter.

If these different scenarios seem confusing, just remember that the fundamental property of a wave is its period. A wave's wavelength, speed, and height do change as it approaches a beach, but its period remains constant.

Airy was an able astronomer, but he was cautious to the point of paralysis. Eventually, that landed him in trouble. He was severely criticized for bungling the opportunity of discovering the planet Neptune. Therein lies an interesting story.

In 1845 John Adams, a British astronomer, deduced the existence of a new planet by analyzing small variations in the orbit of Uranus, and he tried to interest Airy in looking for it. But Airy spent an excessive amount of time checking the accuracy of Adams's calculations. Finally, Airy persuaded astronomer James Challis to search for a new planet, but Challis was unenthusiastic and he procrastinated. Meanwhile, Urban Le Verrier in France predicted the orbit of the unknown planet and was able to inform the Berlin Observatory where to look for it. Berlin found it and snatched the prize from the Brits.

The British scientific community was outraged. Airy defended Adams's priority to the discovery, but the Royal Society (to its eternal credit) awarded the prestigious Copley Prize to Le Verrier. Airy was severely criticized by the British press although he retained his position as Astronomer Royal for another 35 years. And a good thing too, because he accomplished some of his best work during this time. We will meet many scientists in later chapters who have based their theories and models on Airy's fundamental mathematical analysis of weak gravity waves.

Ripples: The Shortest Waves of All

On one of those stifling days of summer, when there is no wind at all, the sea can look as flat as a mirror. Then with the first puff of wind, the surface becomes disturbed. A pattern of very short waves may appear momentarily. Sailors call them "cat's paws." These are capillary waves, better known as *ripples*, which behave differently from gravity waves. We need to understand them because they are the first stage in the growth of ocean waves.

Ripples depend on surface tension, not gravity, as a restoring force. Surface tension, you may recall, is the tendency of the water's surface to behave like an elastic skin that resists stretching. It is caused by the mutual attraction of water molecules at the surface where air contacts water. When the surface is distorted somehow, surface tension acts to smooth it out again. The more the surface is distorted, the more energy it can store. Therein lies the potential for an oscillation and therefore the creation of a propagating wave.

William Thomson (later known as Lord Kelvin) was the first to investigate the properties of such waves. Kelvin was a towering figure in nineteenth-century physics. He made important contributions in several fields of science, including thermodynamics and hydrodynamics. Among other things, he invented a temperature scale with an absolute zero point, at which all molecular motion ceases (the Kelvin scale has its zero at –273°C), and he predicted (incorrectly, as it turned out) the age of the sun. He was also a practical engineer who supervised the installation of the first transatlantic telephone cable. He was brilliant but also rather dismissive of other scientists, especially biologists. He was quoted as proclaiming, "In science, there is only physics—all the rest is stamp collecting."

In the 1870s, Kelvin turned his attention to the physics of water waves. He predicted that the longer the wavelength of a capillary wave, the *slower* it propagates. This is exactly the opposite of the behavior of gravity waves, which move faster the longer the wavelength. To take an example, capillaries with wavelengths of 0.1 and 0.5 cm move at speeds of 68 and 30 cm/s, respectively, and have periods of 0.0015 and 0.016 seconds. Like any type of wave, the phase speed of a capillary is equal to the wavelength divided by the period.

However, because surface tension and gravity act simultaneously on water, an ocean wave may behave either like a capillary wave or a gravity wave, depending on its wavelength. Waves shorter than about a centimeter behave like capillary waves: the longer the wavelength, the slower the phase speed. Waves longer than a few centimeters behave more like gravity waves: the longer the wavelength, the faster the speed. At a wavelength of 1.73 cm, surface tension and gravity are equal in strength, and such an ocean wave moves at a minimum possible speed of 23.2 cm/s, or 0.8 km/h.

Therefore, if we start from a calm sea, we can see how a light breeze creates ripples with millimeter wavelengths that move at very high speeds. As the wind picks up, the ripple speed decreases, and ripple wavelengths increase to centimeters. At a higher wind speed, the wave speed falls to that minimum of 23.2 cm/s and begins to rise again, along with a steady increase in wavelength and in height. The sea now consists of gravity waves.

Kelvin was an excellent scientist, but as he grew older he grew more resistant to new ideas. So, for example, he doubted that atoms exist, he refused to accept evidence for radioactivity and X-rays, and he fiercely ridiculed Darwin's ideas on evolution. We will meet him again in later chapters.

This is a good time to recall the properties of the waves we shall be dealing with. In figure 1.6 we saw waves arranged according to their characteristic periods, from ephemeral ripples to the great ocean tides. Different forces govern these waves. Surface tension and gravity act as restoring forces; wind and submarine earthquakes act as generating forces.

Surfer's Delight

Many surfers share a belief that a swell often contains a "set" of an odd number of waves (usually three or five) with larger-than-average heights. So they wait in the water for them to arrive. These desirable waves have an interesting origin, which I now discuss. We'll also look into the claim that wave sets contain an odd number of big waves.

The swell we see near the shore consists of the remains of the wind-driven waves of a storm far out at sea. The storm produced gravity waves with many different wavelengths. Waves with nearly the same speed obviously tended to travel together, and occasionally they interfered constructively to form a chain of "groups" (fig. 2.4). Each group contains a number of high peaks and is separated from its neighbors by deep troughs. Measurements of swells have shown that a group contains between 3 and 15 wavelengths, with the shorter waves in front and back, and the longer waves in the middle of the group. This pattern of groups is indeed the surfer's desired set, and it has some unique qualities.

Figure 2.4 shows a simple example. In the top graph we see a chain of groups that might represent a swell. Each group has four peaks. A point on one of the constituent waves is marked with a dark circle, and a point on a particular peak of the group is marked with an open circle. The lower graph shows

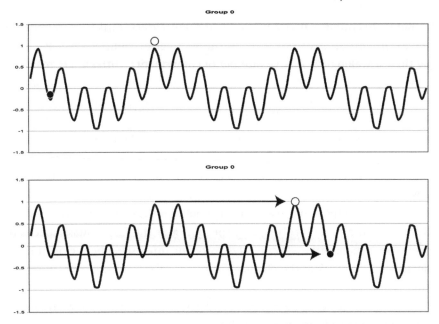

Fig. 2.4 Group speed. *Top*, a wave group (or beat pattern) is formed by the addition of two overlapping waves. A dark dot marks a position on one of the waves, and an open dot marks the tallest peak in the group. *Bottom*, the arrows show how far the wave and the group move in the same time. In general, the group pattern moves at half the speed of the primary waves.

the pattern some time later. The point on the wave (marked by the dark circle) has advanced at its phase speed to its new location. However, you can see that the peak marked with an open circle has advanced only *half as far* as the dark circle. That means that the *pattern* of peaks and troughs travels at a *group speed* which is only half the phase speed of the primary wave.

As a general rule, groups in deep water move at half the average phase speeds of their component waves. In a realistic situation at sea, the short waves exit the group at the rear while the long waves move through the group and exit at the front.

You can also see this effect when you drop a rock into a pond. As the group moves out slowly in a circular wave, the longer waves travel faster than the group, but they die out at the front edge of the group. The short waves travel slower and die out as they emerge from the trailing edge of the group. The total number of waves in the group remains the same.

But is this number necessarily odd, as some surfers think? The number of waves in a group is determined by the number of component waves and by their wavelengths. A large number of components, with closely matched wavelengths, will result in a large number of big crests. These factors are essentially random, however. Therefore, sets with even or odd numbers of big waves are equally probable. Moreover, the shape of the ocean bottom near the shore has an additional randomizing effect on the number of waves in a set. So it doesn't pay a surfer to count the crests in a set and wait for the very biggest one. I suspect that surfers rely more on intuition than on science anyway.

Groups are important because it turns out that the energy carried by waves flows toward a beach at the group speed, not at the faster phase speeds of their component waves. We'll delve into this interesting behavior in more detail when we look at the characteristics of waves resulting from storms in later chapters.

That's probably enough of a meal to digest for now, so let's move on to the question of how a wave is generated in the first place.

CHAPTER THREE

How the Wind Generates Ocean Waves

At first blush there seems to be nothing mysterious about the way wind generates ocean waves. Ask a bright high school student, and he will tell you how it happens. The wind pushes little dimples in the water, and little waves run away from them. These little waves give the wind a better grip on the water and enable it to push the waves harder. Pushing on the little waves makes them grow larger, and pretty soon you have big ocean waves.

That's not a bad explanation, but it skips over a lot of puzzling details. Why doesn't the wind flatten the little waves instead of making them bigger? How does the wind "get a better grip on the water"? And if the wind blew harder, would it cause the waves to run faster, instead of making them taller? And so on.

Physicists have asked such questions about ocean waves for many years. They are motivated partly by pure curiosity, a desire to understand precisely how and why waves grow. But in addition, such detailed information would allow them to forecast the heights of storm waves, helping mariners to avoid the worst of a bad situation.

Over the past century, physicists have devised theoretical models and simulated waves in a computer. They have carried out experiments in the laboratory and at sea. And even so, to this day they still lack a detailed explanation of how ocean waves are spawned and amplified by the wind. Practical forecasting methods have been developed, as we will see, but they are by no means perfect.

In 1879, in his book *Hydrodynamics*, Sir Horace Lamb wrote, "Owing to the irregular character of a wind blowing over a roughened surface, it is not easy to give more than a general explanation of the manner in which it generates and maintains waves." His comment is pertinent even today. But all is not lost, as we will see.

Waves in a Teacup

Lord Kelvin offered one of the first explanations for the birth of ocean waves in 1871. We can picture him in his sumptuous library, blowing on his tea to cool it and beginning to wonder how the little ripples get started. You'll recall that he had previously worked out how the speeds of freely propagating ripples vary with wavelength. Now, Kelvin asked himself, with a wind blowing, which wavelength would appear first and how would that depend on the wind speed? Under what conditions could the wind amplify the wave?

Kelvin attacked the problem from the standpoint of stability. A flat sea, he thought, would remain perfectly flat (that is, stable) until the wind speed reached some threshold, at which point the sea surface would suddenly become unstable and ripples of some undetermined wavelength would spring up. How might this happen?

He was familiar with many situations in which a system becomes unstable. For example, he could picture his teacup resting on a table. A small, steady push at its top edge might tilt the cup slightly but might be insufficient to tip it over. When the force is removed, the cup would return to its original stable position. With a sufficient force, however, the cup could tilt on its bottom edge and flip over. Might something similar occur on the ocean surface?

To test the idea, Kelvin set up a simple conceptual model. He imagined a steady wind blowing smoothly over a flat ocean. Surface tension would tend to smooth out any small disturbance, but changes in air pressure at the surface would tend to amplify it. So he included both forces in his calculation.

To get started, he postulated that a sine wave of infinitesimally tiny height and undetermined wavelength is already traveling freely over the surface. (This was not cheating; it is a standard procedure to introduce a small disturbance: even Archimedes used this method to calculate the area of odd shapes—a pseudo integral calculus method.) What would happen to this disturbance as the wind blew harder?

Kelvin set up the equations of motion and found a solution. It showed that each sine wave begins to grow in height at its own unique wind speed. This critical wind speed is proportional to the wave's free-running speed, its so-called phase speed. Therefore, the sine wave with the slowest phase speed will be the first to grow, and as Kelvin had demonstrated earlier, a capillary with a wavelength of 1.73 cm has the slowest phase speed, 23.2 cm/s.

So far, so good. But Kelvin learned that to create a wave from scratch, the wind must blow much faster than 23 cm/s to compensate for the larger density of water compared to air. In fact, he determined that the wind must blow faster by a factor equal to the square root of the ratio of water and air density, a factor of 27.8. That meant that for ripples with a wavelength of 1.73 cm to appear, the minimum wind speed was precisely 23.2 × 27.8 = 646 cm/s, or 14.4 mph.

This result should have raised a red flag to Kelvin. Surely he knew, as an avid sailor, that these first tiny cat's paw waves appear at wind speeds of far less than 14 mph. So Kelvin's theory failed to match observations. More careful observations showed that 1-cm-wavelength ripples appear at a wind speed less than 1 km/h (0.6 mph).

Was Kelvin's theory wrong? No, his theory of instability was correct, as later experiments showed, but certain factors he chose to neglect in his model, such as wind turbulence, are more effective in raising a water wave at lower wind speeds. (I will examine the role of turbulence in a moment.) Kelvin had toyed with the idea of taking into account the friction between air and water, which he had neglected so far, but he did not have the mathematical tools needed to describe such friction or other turbulence. As we will see, Kelvin's neglect of turbulence in the wind turned out to be the crucial flaw in his theory. He had, in effect, made the problem more difficult by assuming the sea remains *perfectly flat* until a critical wind speed is reached.

But let us give him due credit; he had a clear idea of the forces at play in generating waves. Here is a heuristic explanation of his mechanism as Kelvin might have imagined it. In figure 3.1 we see a very weak wave with some arbitrary wavelength. As the wind flows smoothly and steadily over the crest of a weak wave, centrifugal force pulls the air upward and so reduces the pressure there. (Recall that centrifugal force acts on a body when it is forced to travel in a curve. Anyone who has enjoyed a ride on a roller coaster has felt the centrifugal force that tends to throw one upward at the top of the ride). The decrease in pressure tends to suck up the crest slightly. In the water troughs, the centrifugal force is directed downward, just as you are pushed down in your seat at the bottom of a roller coaster "valley." Here, the increase in pressure tends to deepen the trough.

At low wind speeds, this force is resisted by the surface tension of the water. But the centrifugal force increases as the square of the wind speed. When the

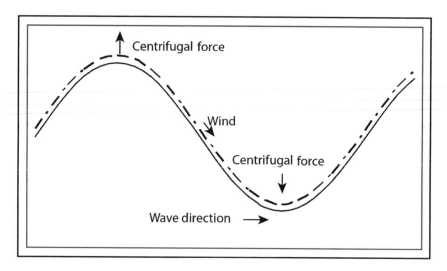

Fig. 3.1 The Kelvin instability. As wind flows over a water wave, centrifugal forces at the crest and trough act to increase the height of the wave. At a critical wind speed these forces disrupt the wave.

wind speed reaches the critical speed for this wavelength, the centrifugal force pulling up on the crest exactly equals the weight of the water. Kelvin would conclude that any wind faster than critical would cause the height of the wave to increase rapidly. The wave would become unstable.

In studying the problem, Kelvin discovered a fundamental instability at the surface between any two flowing fluids, an effect that has wide applications in science. His German colleague Hermann von Helmholtz discovered the same effect independently, and the instability is named for both of them. The instability is seen in many situations: in billowing clouds, in cigarette smoke as it wafts into the air, and in astrophysical plasmas.

Pushes and Pulls

After Kelvin's attempt, progress in explaining the generation of ocean waves stalled for over 50 years. Then in 1924 Sir Harold Jeffreys, the dean of geophysicists in the early twentieth century, published a fresh idea. It concerned the growth of an existing ripple, not its birth. Basically, he proposed that the wind, moving in the same direction as a wave but somewhat faster, would tend to separate from the water surface as it flowed over a crest. In effect, the wind would leap from crest to crest, avoiding the troughs. The back side of the crest would then *shelter* the front side of the crest, so the pressure on the front side would be lower. As a result, the wind would exert more horizontal pressure on a crest's back side than on its front face.

The difference in pressure from back to front would exert a net horizontal force, which would tend to accelerate the wave. But the wave's phase speed is fixed by its period (or wavelength), assuming that Airy's theory applies. Therefore, the wave would absorb the energy input from the wind by increasing its wave height.

Jeffreys's mechanism (called "form drag") was so appealing that a number of laboratory experiments were carried out to test it. Unfortunately, they showed that Jeffreys's predicted pressure difference was too small to account for the observed rate of growth of wave heights. Another good idea shot down!

But Jeffreys had many scientific successes to compensate for this failure. In 1923, for example, he predicted that the surface temperatures of the outer planets were well below zero, not red-hot, as most scientists of the time believed. He was proved correct much later. On the other hand, he firmly rejected the concept that continents could drift. Alfred Wegner had proposed the idea after pointing out how the coasts of South America and Africa could have fit together in the past. Jeffreys thought that no geological forces were strong enough to shift a continent. Only in the 1960s, with the discovery of sea-floor spreading and tectonic plates, was Wegner proven correct. In such ways can scientists be both geniuses and misguided.

Joseph Fourier

Before we continue with this story, we need to take a small detour and talk about Fourier methods. Joseph Fourier was a French mathematician and physicist who made important contributions to the theory of heat in the nineteenth century. In 1822 he introduced an idea that has enormous utility in all sorts of applications. He proved that any reasonably smooth curve could be represented as a sum of sine and cosine waves of different wavelengths and amplitudes. Moreover, he showed exactly how this could be done mathematically. The different sine and cosine waves are chosen to overlap constructively at some places and destructively elsewhere. In this way the peaks and valleys in any given curve or surface can be fitted, more and more accurately, by adding more and more trigonometric waves.

Figure 3.2 shows an example of this process. The dark, irregular curve called "combined wave" might represent the crests and troughs of a choppy

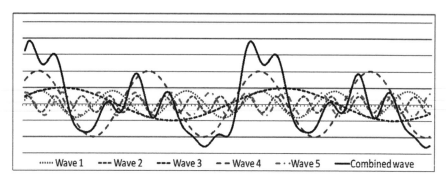

Fig. 3.2 An example of fitting a given curve with sine and cosine curves, following the recipe of Joseph Fourier.

sea. Notice the high maximums at the left side and middle of the figure. The different curves you see under the maximums are different components of the combined wave. They are sine waves of different wavelengths. If the sum of, say, 20 different sine and cosine waves were used to represent the original curve, we'd see how closely the dark curve could be reproduced.

Now let's resume our story.

Resonance Enters the Picture

Further progress in solving the wave generation problem was delayed for another three decades. Then, in the same year (1957) two men, separated by the Atlantic Ocean, published independent and quite different solutions to the wave problem. Owen M. Phillips was working at St. James College, Cambridge, and John W. Miles was at the University of California at Los Angeles.

Phillips's Resonance Model

Owen Phillips was Australian by birth but spent most of his career at Johns Hopkins University in Baltimore. He was the pupil of two of the foremost fluid dynamicists of the twentieth century, G. I. Taylor and G. K. Batchelor, both at Cambridge University. In 1955 Phillips completed a doctoral thesis on the effect of turbulence on aircraft wings and then shifted his attention to hydrodynamics. Eventually, he decided to move to the United States, where oceanography was progressing more rapidly. He had wide interests in fluid dynamics, including the flow of magma in the interior of the earth. In 1965, Phillips was awarded the coveted Adams Prize by the Royal Society of London for his first monograph, *Dynamics of the Upper Ocean*. He was professionally active until his death in 2010 at the age of 79. We'll encounter his work in several chapters of this book.

Phillips introduced the idea that turbulent pressure fluctuations in the wind might generate ripples if certain conditions for resonance were met. *Resonance* refers to the agreement between a system's natural frequency and the frequency of an oscillating force. A child's swing is a good example. If the swing is pushed at the natural frequency of the swing, it swings wider and wider. Likewise, a soprano can shatter a wineglass if she sings the note that corresponds to the natural frequency of the glass.

Phillips visualized an initial situation in which a smooth, steady wind is flowing over a flat sea. From his studies of turbulence on aircraft wings, he knew that the wind does not remain smooth and uniform for very long. Even the least friction between the air and water causes turbulent eddies to form in the air. Eddies of all sizes and lifetimes develop rapidly and travel downwind.

Phillips knew that the small, turbulent eddies in contact with the water surface exert random pulses of pressure on the surface (fig. 3.3). The water under an eddy responds by beginning to oscillate in height. The periods of oscillation vary randomly from one eddy to another along the surface, however. It would seem that there was no obvious means of converting these local oscillations into organized wave motions.

But Phillips had a brilliant idea. Suppose the whole distribution of pressures along the sea surface changed rather slowly (during the period of, say, a ripple) and just drifted intact, downwind at the speed of the wind. Then he could imagine deconstructing this "frozen" and jagged pressure distribution into sine and cosine waves of different wavelengths, using Fourier's method.

Next, he imagined a very weak water wave traveling at an angle to the wind (fig. 3.3, bottom). If a pressure sine wave, when projected on the direction of the water wave, has the same speed and wavelength as the water wave, then the two waves overlap and are in resonance. The air pressure pushes down on the water troughs and lifts up on the crests at exactly the right moments. Therefore, the pressure wave can amplify the water wave or even generate it from scratch. Each pressure sine wave might resonate with a water wave, and so a variety of water waves could be generated.

Phillips worked out the details of his theory. His main result was that the distribution of energy among water waves of different wavelengths depends

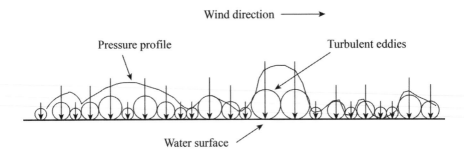

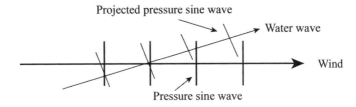

Fig. 3.3 Phillips's model of wave generation. Turbulent eddies of air near the water surface (*circles*) exert random pressures on the water. The pressure distribution is assumed to drift downwind intact at the wind speed. If the pressure distribution happens to match the sinusoidal profile of some arbitrary freely running water wave, as in the lower graph, the two sine waves will be in resonance and the pressures will generate a wave.

on three factors: the pressure distribution at the sea surface, the length of time that the pressure distribution at the sea remains unchanged, and the *square* of the ratio of densities of water and air. This last factor is very small, about one-millionth, and sets the scale of the effect that Phillips predicted. Based on this analysis, he also calculated that the wind energy would be transferred to water waves at a constant rate, independent of the heights of the waves. However, as we will see, this analysis was at variance with other theories.

At the time Phillips wrote, there were no observations of air turbulence at the sea surface, and only a few over a land surface. Using the best data he could find, Phillips predicted that the first waves to appear would have the minimum possible wavelength that Kelvin had predicted (1.7 cm). But unlike Kelvin, Phillips predicted they would appear at a wind speed of 23 cm/s (less than 1 km/h), not 646 cm/s. To that extent, Phillips's theory was in far better agreement with what every sailor knew.

Here then was a mechanism for generating the first ripples by means of random pressure fluctuations. Phillips's theory depends critically, however, on the assumption that the random distribution of turbulent pressure along the surface of the sea will persist and simply propagate downstream intact. But the small eddies which are most effective are just the ones that change most rapidly. And the overall efficiency of the process is small because of the tiny ratio of air and water densities. Nevertheless, Phillips's mechanism is still invoked in some prediction schemes to begin the generation of waves, as we will see.

Miles's Resonance Model

John W. Miles was unaware of Phillips's theory when he published his own theory in 1957. At the time, Miles was a professor in the engineering department of the University of California at Los Angeles. During World War II he had worked on radar at the Massachusetts Institute of Technology and on turbulence at the Lockheed Aircraft Corporation. After the war he was appointed to a professorship at UCLA, where he remained until 1964, when he moved to the Scripps Institution of Oceanography. There, he explored a wide range of subjects, including ocean tides, the stability of currents, and the interactions of water waves. Like Phillips, he had worked on the development of turbulence and vortices on aircraft wings, and these studies may have led him to the problem of how ocean waves originate.

Miles set out to explain the *growth* of water waves, not their initial generation. Like Phillips, Miles proposed that air pressure fluctuations near the water surface amplify waves. Miles, however, introduced the novel idea that once weak waves appear, they modify the airflow and therefore the pressure distribution near the water surface, in such a way as to amplify themselves. It is a resonance or positive feedback scheme, a bootstrap operation.

Miles began by adopting a realistic wind profile in which the air speed is small near the sea and increases steeply at greater heights. Although this classical profile is the result of the turbulence that normally arises between layers of air that move at different speeds, Miles ignored any effect turbulence might have on the growth of waves. In effect he assumed that layers of air slide past each other without interacting, a so-called laminar flow. This was a weak point in his theory that he addressed later on.

He imagined that a weak sinusoidal gravity wave is already propagating downwind at its characteristic phase speed. As the wind flows smoothly over the surface of the wave, the air pressure decreases over the crests and increases over the troughs, as in figure 3.1. These pressure changes propagate upward as a sound wave into the region of increasing wind speed (fig. 3.4). At a critical height, where the wind speed equals the phase speed of this water wave, this pressure wave draws energy from the wind and is reflected downward.

We'll bypass the rather subtle details of the dynamics at this critical height. It's enough to say that reflection of the pressure wave at the critical layer causes the pressure distribution at the water surface to shift in an interesting way.

A slight increase in pressure is exerted downward, a quarter of a wavelength back from the water wave's crest, where the water is already moving downward (fig. 3.4). On the front of the crest, where the water is already moving up, a slight decrease in pressure acts to enhance the upward movement of the water. The combination of pushes and pulls delivers wind energy to the wave. The water wave also begins to change shape and tip forward. Miles's result may remind us of Sir Harold Jeffreys' "sheltering" theory, in which the air pressures differ from front to back of a crest; but it differs in the important positive feedback role of the critical layer.

Miles was able to estimate the rate at which a water wave of an arbitrary wavelength gains energy. He found that the rate of growth of wave energy is

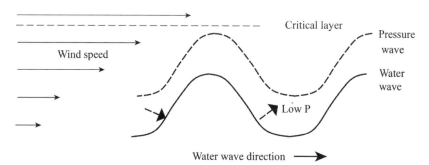

Fig. 3.4 Miles's model of wave generation. The wind speed is assumed to increase with increasing altitude, as observations have shown. Near the water surface the wind follows the profile of a weak water wave; above the surface a pressure wave is set up in the air. This air wave interacts with the critical layer of air, where the wind speed equals the phase speed of the water wave. The interaction produces weak pressures (*heavy arrows*) at the front and back of the water wave's crest that act to amplify the wave.

proportional to its present energy. Therefore, the waves will gain energy at an ever-increasing (exponential) rate, not at a constant rate, as in Phillips's theory. Moreover, the growth rate is directly proportional to the air-water density ratio and not to the square of that ratio, as in Phillips's theory. (Remember that this ratio is about 0.001, so that the square is much smaller, about 0.000001.) Thus, Miles's mechanism is much more efficient.

Miles's theory was met with a lot of skepticism soon after it was published. The most serious criticism was that the theory ignored turbulence altogether, despite the fact that the realistic wind profile that Miles adopted is maintained by turbulence. The critics argued that his wave-induced pressure waves might be swamped by turbulence in the wind.

Moreover, it was argued that Miles had grossly simplified the mechanism that transfers energy from the wind to the wave at the critical height. And he had also ignored the possibility that the growing water waves would change the assumed profile of wind speed. Finally, Miles ignored the drag of the wind on the water surface and focused only on pressures perpendicular to the surface. All of these weak points in Miles's theory were explored by theorists in the decades that followed. Even after much research and some improvements by other theorists, Miles's theory was still considered controversial as late as the 1990s, when improved observational techniques were able to vindicate him.

A Test of Theory at Sea

The real test of a theory is a careful experiment. When Miles's predicted growth rates were compared with observations taken in the ocean in 1971, they were found to be too small by at least a factor of 10! Not exactly a brilliant success.

However, observations of such growth rates are technically very difficult and are therefore subject to bias. In 1974 R. L. Snyder, of Nova University, decided to take a different approach. He and his colleagues resolved to measure just the wave-induced air pressure that was so important in the Miles theory and the corresponding transfer of wind energy to the waves. To this end, they set up an experiment in the Bight of Abaco, Bahamas, in November and December of 1974. They deployed a three-dimensional array of six wave pressure sensors, seven air pressure sensors, and a device that measures wave heights. They obtained 30 hours of high-quality data.

Their results on wave growth were consistent with Miles's theory but only approximately. The theory predicted growth rates for waves (especially long waves) that were too small by a factor of at least 2 to 3. Snyder and friends repeated the experiment in 1981, with no great improvement in the agreement between theory and observations.

Waves in the Lab

Meanwhile, other researchers in the United States and Japan were exploring the growth of waves in the laboratory. Hisashi Mitsuyasu was studying capillary waves at Kyushu University. E. J. Plate was examining the growth of micron-amplitude capillaries at Colorado State University, while William J. Plant and John W. Wright were investigating short gravity waves at the U.S. Naval Research Laboratory, in Washington, D.C. In the late 1960s and early 1970s, each group built a sophisticated machine: a wave tank enclosed in a wind tunnel. It would enable them to measure the growth and equilibrium of waves under carefully controlled conditions.

We'll focus on the experiment of Plant and Wright. They were interested in deep-water gravity waves between 4 and 36 centimeters in length. They could turn on a wind of a chosen speed and watch it raise waves on the flat water in the tank. After a while, the waves would settle into a steady pattern. Then, using a clever Doppler radar technique (which I discuss later), they could measure the average height and speed of selected wavelengths at several distances along the length of the tank. Their work gives us a better picture of a wave system in equilibrium with the wind.

They saw that when a steady wind blows over the water for a while, the waves grow to a maximum height and stop growing. The longer the distance that the wind blows without interruption (its "fetch"), the higher the waves rise, as you might expect. In this steady state, the energy that a wave absorbs from the wind must be balanced by the energy it loses by various dissipative forces, such as wave breaking or turbulence.

You might think that a steady wind would raise waves of a single preferred wavelength, as in a wave tank with an oscillating paddle. That is not what Plant and Wright saw, however. Instead, at each distance along the tank they detected a narrow range of wavelengths, centered on a "dominant" wavelength. This dominant wavelength was longer at longer fetches and also longer at higher wind speeds.

So, for example, at a wind speed of 7.6 m/s, the dominant wavelengths at fetches of 2 m and 10 m were 6 cm and 24 cm, respectively. With a steady wind of 11 m/s, the dominant wavelengths increased to 12 cm and 38 cm at these same fetches. We'll see the same effect in a famous experiment carried out at sea.

At the beginning of a trial run in the lab, when the wind was turned on, the waves grew exponentially in strength, just as Miles had predicted. But the measured growth rates of the longer waves were significantly larger than theory predicted, just as Snyder had found in his Bahamas tests. The reason, the experimenters suggested, was that energy was being transferred from short to long wavelengths, an effect that Miles had not anticipated. That explanation could also account for the shift to longer dominant wavelengths at longer fetches. This energy transfer effect, an important factor in wave growth, was first predicted in 1962 by Klaus Hasselmann, whom you will get to know quite well in later chapters.

In 1989 Peter Janssen, a theorist at the Netherlands Meteorological Institute, improved Miles's theory significantly. Janssen recognized that as waves grow at the water surface, they react back on the wind and change the wind speeds at different heights. Miles had assumed the wind profile remains constant. Janssen went on to calculate the steady-state wind profile that results from this interaction. His work has been incorporated in modern methods of wave forecasting, as we will see.

Another Test of Miles's Theory

Snyder's experiments at sea had failed to detect the critical layer that plays such an important role in Miles's theory. Finally, in 1995, Tihomir Hristov repeated Snyder's experiment with improved techniques of analysis and was able to find the elusive critical layer. His result lent strong empirical support to Miles's theory.

Hristov, a Bulgarian by birth, is a theorist at Johns Hopkins University. He originally studied turbulent flows at the University of Sophia; then, for his postdoctoral research he moved to the University of California at Irvine. There, he met Professor Carl Friehe, who interested Hristov in the vexing problem of wave generation.

After studying Miles's theory, Hristov developed an analytical method to extract the critical wave-induced air pressures from the much larger turbulent pressures in the wind. As a result, he thought he could make a rigorous observational test of Miles's theory. But he needed the help of some experienced experimentalists.

So in 1995 he and his colleagues S. D. Miller and Carl Friehe set up their experiment 50 km off the coast of Monterey, California. They were working on a specialized ship, the Floating Instrument Platform, or FLIP, that the U.S. Navy had built in the 1960s. By flooding the onboard tanks, the crew could rotate the FLIP from a horizontal to a vertical position in the sea. Just like an iceberg, most of its 108-m length was submerged; this prevented the FLIP from bobbing up and down even in a rough sea. The FLIP was so stable that the surface elevation of the sea surface could be measured with an accuracy of one centimeter.

The team installed a stack of 12 anemometers on a vertical mast, which was mounted on a boom at some distance from the FLIP. These instruments measured the wind speed and direction at different heights above the sea. In addition, rapid variations of wind speed were recorded at 50 Hz with special anemometers at four heights. Finally, continuous measurements were made of the elevation of the sea under the vertical mast.

During the five days of the experiment, the wind speed varied from calm to stormy (15 m/s—or about 32 mph). After the data were collected, Hristov extracted the pattern of the wave-induced air flow. It confirmed Miles's theory in considerable detail. In particular, the theory correctly predicted how, at the water surface, wave-induced pressures shift backward by a quarter of a wavelength relative to their parent water wave. Hristov and company wrote: "Throughout the experiment the wave-induced flow maintains the critical layer pattern. The numerical and the experimental results seem to agree closely." Miles, then in his mid-70s, must have been enormously pleased.

The main point of Miles's theory is that the wind and the sea are tightly coupled in the layer near the surface: the wind generates the waves, and the waves, in turn, affect the wind profile. With Janssen's modifications, Miles's theory is now considered a central part of modern numerical models of ocean wave generation. However, as we will see in chapter 5, some oceanographers regard Miles's theory as still not quite complete and believe it has never been fully tested. They continue to examine the intricate processes at the air-water surface.

That Puzzling Critical Layer

John Miles was rather coy about explaining what exactly goes on in the critical layer of his theory. When Hristov asked him about it during a 1998 seminar, Miles smiled and replied it was a "mathematical convenience." But this critical layer is the key to the transfer of energy from the wind to a weak water wave, and a fuller explanation was needed.

Michael James Lighthill, a British physicist, actually provided one in 1962, and his ideas were further developed by Soviet scientist V. P. Reutov in 1980. In this expanded model, the critical layer corresponding to each wavelength consists of a row of vortices. Each vortex rotates about a horizontal axis, with its rotational speed slowest at the axis. Lighthill realized that because the air just above the critical layer is flowing faster than the water wave, it can be a source of energy. Some of this fast air is trapped momentarily in the critical layer's vortices, is spun up by the vortices, and then exits below the layer. The vortices transfer their excess rotational energy down to the water surface. This mechanism seems to work despite the near absence of viscosity, which would help to couple different layers of air.

Another Idea

Measuring the growth rate of waves at sea is difficult, and observations of rates over the past 40 years scatter by at least a factor of 2. The experimental situation is unsatisfactory. But could it be that some important physical factor has been overlooked?

In 2008 Brian Farrell (Harvard University) and Petros Ioannou (University of Athens) assumed that all the measurements of growth rates are of equal value but that some unknown factor was causing the spread. They then suggested that *gusty* winds during the observations might have caused the scatter in the data. They also noted that Miles's theory assumes a steady wind. So the researchers performed numerical simulations to evaluate the effect of gusty winds on wave growth rates. They were able to show that gustiness can explain much of the scatter. The Kelvin-Helmholtz instability, discarded long ago, seems to work nicely when gusty winds are included. But do the experimenters agree? This new idea will no doubt be examined carefully. Stay tuned!

At this point you may be wondering why the generation of ocean waves seems so complicated and is so important to get just right. Whatever happened to the appealing and simple idea that the wind *rubs* on the flat surface of the water and makes ripples, which grow larger as the wind pushes them? The short answer is that the behavior of the layer of air in contact with the water surface is extremely complicated when examined in any detail and yet has a profound effect on the types and energies of waves.

When the wind blows over the flat sea, Kelvin instability occurs, and the layer becomes turbulent. Transient vortices are set up, and the surface becomes roughened. Many theorists have tried to frame a mathematical description of the process, moment by moment, but without complete success.

But considerable progress has been made, with a combination of wind-wave tank experiments and elaborate numerical simulations. For example, the Taiwanese researcher M. Y. Lin and his associates have modeled the two-dimensional evolution of small waves in the turbulent air-water interface. Their results, published in 2008, are fascinating but too complex to summarize here. We can only say that the subject is still in a state of flux.

In this chapter I've been talking mostly about the generation of weak waves or waves in the early stages of growth. In the next chapter we examine more realistic waves that have grown to finite height.

CHAPTER FOUR

A Touch of Reality

How Big Waves Behave

If you have ever been at sea during a gale, you will never forget the awesome power of the great wind-driven waves. They rise majestically, foaming and spitting spume. They sweep by, pitching you into the deep troughs. Each one seems taller than the next, and each one threatens to overwhelm your puny ship.

Mariners often face such conditions in a voyage. From time immemorial, they have wanted reliable predictions of the waves they were likely to encounter. An accurate forecast could warn them to avoid the center of a storm. It could mean the difference between surviving and capsizing.

Oceanographers have worked hard to satisfy this demand, but it was not until the mid-twentieth century that they were able to provide some relief. To reach the final goal of a reliable forecast, they first had to learn how to measure storm waves and to describe them mathematically. And they had to understand how big waves differ in behavior from the very weak waves described by Airy. In this chapter we follow a few threads of their research.

The Challenge of Describing Ocean Waves

Take a look at the stormy sea in figure 1.5. Waves seem to be arriving from several directions, with different heights and wavelengths. In a few moments the scene will change dramatically. How could anyone describe such a scene quantitatively?

Oceanographers have devised a practical approach to this challenge. They take a census of the waves—that is, sort them according to their wavelengths, directions, and heights. But they don't attempt to track individual storm waves; that would be well-nigh impossible. Instead they describe the sea *statistically*. Their idea is to measure the heights of waves at one or more fixed positions in the sea as a function of time and then analyze the data to produce a so-called *wave energy spectrum*. A spectrum is a graph that shows how energy is distributed among waves of different periods or wavelengths. It can be compared with a predictive theory if one is available.

A rainbow is probably the most familiar example of a spectrum. When sunlight passes through a veil of water droplets, we see the colors of the rainbow spread out in a spectrum. Each color (that is, wavelength) has a definite amount of energy (brightness). We could measure the energy at each color with a photocell, record the results, and plot them in a graph. This would be the quantitative spectrum of sunlight. If we repeated the procedure with, say, fluorescent light, we'd learn that it contains a different mixture of wavelengths and therefore a different spectrum.

We can understand the concept more easily if we look at a specific example. Let's imagine an oceanographer (let's call her Dr. O) who wants to measure the energy spectrum of ocean waves. She will measure the heights of waves at a fixed location in the sea by measuring the pressure beneath them, using a pressure sensor. The higher the wave, the larger is the pressure under the crest. She will record the data for some length of time and then use a mathematical tool, devised by the French physicist Joseph Fourier (whom we met earlier), to calculate the energy spectrum of the waves.

Let's imagine that Dr. O attaches an underwater pressure sensor on the leg of one of those stationary offshore oil rigs. The sensor measures the varying pressure under waves as they pass by. A simple calculation then converts the water pressure to a wave height. A record of the varying heights might look like the solid dark curve in figure 3.2.

Now it is very likely that in a stormy sea, several waves are overlapping at the leg of the oil rig at any instant. A few seconds later a different combination of waves may overlap at the leg. That is why the record looks so jagged. Dr. O wants to sort all the waves that are present in the sea according to their periods and determine their individual amplitudes. Then she can determine the energy each wave carries, since it is proportional to the square of the wave's amplitude. And from these data she can plot a graph of energy versus period (or frequency, the inverse of period). This will be her desired spectrum of the stormy ocean. But how can she accomplish this task?

Joseph Fourier, that famous French mathematician and physicist, will come to her rescue. As I described in chapter 3, Fourier proved that any curve or surface, with few exceptions, could be represented as a sum of sine and cosine waves of different wavelengths and amplitudes. Moreover, he showed exactly how this could be done mathematically. The different sine and cosine waves are chosen to overlap constructively at some places and destructively elsewhere. In this way the peaks and depressions in any given curve or surface can be fitted, more and more accurately, by adding more and more trigonometric waves.

Following Fourier's idea, we may view the ocean surface as made up of overlapping sine and cosine waves with a wide range of periods, each moving with its own particular speed, amplitude, and direction. Each sine wave carries an amount of wave energy that is proportional to the square of its amplitude. Our oceanographer wants to know how much energy is carried by waves of different wavelengths or periods.

This may seem like a daunting task, and to simplify the problem Dr. O has limited herself to sampling the heights of waves that pass by one point in space, the leg of the oil rig. If her observations extend over several hours, she can safely assume that every wave with a shorter period than this observation time will appear at her detector and will be recorded. That is the assumption she has to make to obtain a statistical description of the sea.

We'll skip the details of the Fourier analysis of this raw data; Dr. O gives this task to her graduate students to undertake. For the present it's enough to say that with the use of Fourier's method, the slow undulations in Dr. O's digital record (fig. 3.2) can be fitted with long-period sine and cosine waves, and the rapid variations with shorter and shorter–period sine and cosine waves. For each trigonometric wave Dr. O introduces, she can find the total energy present in the sea in waves with a definite period.

Figure 4.1 shows the spectrum she might find—the amount of energy as a function of frequency, which equals 1 divided by the period. Most of the energy is concentrated in a narrow range of frequencies, typically between 0.1 and 0.4 cycles per second (hertz) or periods of 10 and 2.5 seconds.

The spectrum may change during a long storm. In fact, if Dr. O carried out the calculation with only the first half of her digital record, repeated it for the second half, and compared the results, she might very well see how much the spectrum varies in time.

But Dr. O is looking only at one spot and so cannot tell what direction these waves are going. As noted earlier, we can consider the surface of a stormy sea to be a mixture of sine and cosine waves of different periods, but these will be running in many directions away from the center of a storm. If we could take a snapshot of, say, a square kilometer of the sea and could measure the

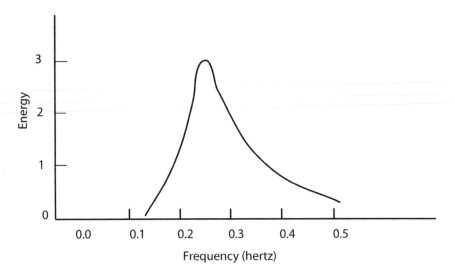

Fig. 4.1 A spectrum shows the distribution of energy among water waves of different frequencies.

instantaneous height of the sea at every point, we could apply Monsieur Fourier's method. We could decompose the surface into sine and cosine waves, along the north-south and east-west directions. With this information we could derive the instantaneous energy spectrum with a good deal of certainty. And the spectrum obtained in this way would certainly be more reliable than the spectrum Dr. O could obtain from a single point in space. In addition, Dr. O could determine how the spectrum varies in different directions. We will see in a later chapter how radar satellites allow us now to take such snapshots and deliver high-quality spectra.

Two Pioneering Applications of Wave Spectra

Oceanographers have determined that the spectra of swells change more slowly than the spectra of waves in the middle of the stormy area. Swells, you will recall, are the trains of long-wavelength waves that propagate away from a storm center. Walter Munk, one of the founders of physical oceanography, used this property in a classic study of the propagation of swells.

As a young man Munk was pointed by his father toward a career in banking. After working in a bank for three years, he rebelled. He quit a job he hated, enrolled at the California Institute of Technology, and obtained a bachelor's degree in physics in 1939. His summer job at the Scripps Institution of Ocean-ography (La Jolla, California) changed his life: he decided on a career in oceanography. Armed with a doctorate in geophysics from the University of California at Los Angeles, he found a job at Scripps in 1947 as an assistant professor of geophysics.

In 1959, Munk and three colleagues from Scripps set up a triangular array of three pressure gauges at a depth of 100 m, 3 km off San Clemente Island, California. Even at these depths, their sensitive sensors could detect the changes in pressure due to the heights of passing waves.

Munk and his associates recorded the variations of pressure continuously for several months. Every day, they computed the energy spectrum of the waves that arrived at each station in the array. The low-frequency (long-period) waves arrived first because their speed is higher, as we learned earlier. The higher-frequency waves arrived later and carried much more energy. From the lag of the higher-frequency waves and from their phase speeds (calculated with Airy's theory), Munk could estimate the distance of a storm along a great circle of the earth. One of these turned out to be a staggering 12,000 km, or 7,500 miles.

From the difference in times of arrival at the three stations in the triangle, the scientists could also determine the direction of the incoming swell. The distance and direction of this source of the storm noted above pointed to the Southern Ocean, between New Zealand and Antarctica. From the details of the data, Munk could also estimate the wind speed that produced the swell. When he compared his deductions with real weather maps, he found good agreement in most cases.

Munk was amazed at how little a swell decayed as it propagated long distances. To find out just how a swell decays, he decided to set up another experiment. In 1963, with the support of the U.S. Navy, he established six recording stations on a great circle from New Zealand to Alaska. Each station measured pressure variations at various depths. Five stations were near islands. The sixth station was that remarkable ship, FLIP, which could be flooded to cause the vessel to rest vertically in the sea. In this position the ship provided a very stable platform from which Munk could measure pressure variations in the water.

Munk learned that the swell decayed very little after it had traveled a distance from the edge of a storm equal to the storm's diameter. The swell also decayed by spreading sideways, not because of the resistance of the air above it, as most oceanographers believed at the time. Apparently a swell is finally absorbed at the boundaries of the ocean, presumably as surf.

Walter Munk is a remarkable man. In a career spanning 72 years, his research interests have covered a very wide range. They include the rotation of the earth, wind-driven circulation in ocean basins, the tides, sonic tomography of the ocean, and ocean swells. In the early 1940s, Munk and Harald Sverdrup practically invented the science of forecasting wave and surf heights, as we will see in a later chapter.

Munk helped to make the Scripps Institution into the world-class center of ocean research it is today. Munk has received a dozen prestigious awards for his many contributions to science, including the National Medal of Science (1977). In 2012 he received the Crafoord Prize by the Swedish Academy of Sciences. It's the nearest thing to a Nobel Prize.

At age 94, Munk married a younger woman. The *San Diego Union-Tribune* of March 23, 2012, quoted him as quipping, "When I came to La Jolla, the saying was that the community consisted of the newly-wed and the nearly dead. I now qualify on both counts."

How High Waves Behave Differently from Low Waves

Not all waves are simple swells. Water waves are rather complicated once one begins to examine them closely. The pioneers of wave theory—men like Lagrange, Laplace, Cauchy, and Kelvin—understood this very well. The wave equations they had to deal with include some nasty "nonlinear" features, such as a product of velocity and the gradient of velocity. That meant that each point on a wave profile could have a different forward speed. The wave profile would evolve in time, which would require horrendous calculations to follow.

To obtain a wave with a stable profile that could be readily modeled mathematically, it was necessary to limit the wave height to a tiny fraction of the wavelength. Only then could one ignore the troublesome nonlinear features. The resulting "linear" theory of weak (low-amplitude) waves, summarized by Airy, has turned out to be very useful, despite its limitations, in getting a rough estimate of the behavior of realistic waves.

But to understand the behavior of real storm waves, which are dangerous and therefore more important to mariners, mathematical physicists have had to develop new theories of waves with heights comparable to, say, a tenth of their wavelengths. These waves turn out to have some unexpected and puzzling features, particularly the ability to exchange energy when they collide. At the same time, oceanographers have had to develop techniques for measuring the chaotic waves one sees in a stormy sea. This dual approach of crafting theory and experiment in parallel has paid off magnificently.

Russell's Solitary Wave

The theory of realistic water waves was launched in 1847 with the publication of a brief technical report by John Scott Russell, a British engineer. Russell had made a startling discovery in 1834 while working on the optimum shape of canal boats. He was watching a boat being pulled slowly on a canal by a team of horses. At some point the boat stopped abruptly, and a large "solitary heap of water" rolled ahead at high speed. Russell chased this wave on horseback. He reported later that the heap was about 1 foot high and 30 feet long and moved at about 9 mph. He was most impressed to see that this solitary wave preserved its smooth, rounded shape for a distance of 2 miles.

Fascinated by this unique wave, Russell carried out experiments in a wave tank at home to investigate others like it. He named these waves "waves of transition." He found that the speed of such solitary waves increases as their height increases, unlike the weak waves Airy described.

His experimental results were first ignored and then contested by the leading theorists of the time because they seemed to conflict with known physics. Airy, for example, calculated that different parts of a tall wave travel at different speeds. The speed of a chosen point on the wave profile depended on the elevation of the point above the undisturbed water level. So, for example, the crest moves faster than the trough. Therefore, Airy concluded that a big wave cannot propagate without a severe change of shape. The slope will become steeper at the front face of the crest and flatter at the back face. The wave will ultimately curl over and collapse. Russell, he concluded, must be mistaken.

Stokes Waves

Sir George Gabriel Stokes picked up the problem upon reading Russell's report. Stokes (1819–1903) was a physicist and mathematician whose research at Cambridge University covered many different fields over a period of 50 years. He was a contemporary of two other Cambridge luminaries, James Clerk Maxwell and Lord Kelvin.

Stokes graduated in 1841 from Pembroke College with the highest honors in mathematics. He was elected to a fellowship, and in 1849 he was appointed as Lucasian professor of mathematics at Cambridge. Stokes was renowned for his extraordinary combination of mathematical power and experimental skill, which gave him the ability to find solutions to seemingly intractable problems. He was interested in all aspects of wave dynamics, including water, light, and sound waves.

In 1847, Stokes decided to try to explain John Russell's wave. He searched for a mathematical solution to the full nonlinear wave equation that would describe a gravity wave of finite height that could propagate without a change of shape in shallow water. He eventually determined that the best solution has the shape of a trochoid, the curve traced in space by a point on the rim of a rolling wheel. Stokes's wave has a sharper peak and a broader trough than a sine wave of the same wavelength. If the wavelength of a Stokes wave is sufficiently large, each crest appears to be a single pulse, a solitary wave like the one Russell observed.

A Stokes wave has some novel properties. Its speed is larger than that of a small-amplitude wave of the same wavelength. And, in agreement with Russell's experiments, the wave speed increases as the wave grows in height. Also, the height of a Stokes wave has a limit: the wave will become unstable and break, even in deep water, if its height exceeds 14% of its wavelength. Here is a possible clue to the origin of breaking waves at sea, the so-called whitecaps.

Stokes also predicted that the blobs of water do not move in closed circles under a trochoidal wave. In one oscillation a blob is carried forward by a small amount in the direction of the wave's propagation, an effect called Stokes drift. Real waves display this behavior. A swimmer resting in the swell soon becomes aware of slowly drifting toward the beach. (In chapter 2, I purposely ignored this small, persistent drift.)

A Stokes wave somehow avoids dissolving into its component sines and cosines as it travels with constant speed. At first sight this seems to be unnatural. As Airy commented, we should expect each wavelength in a composite wave form to travel at its characteristic phase speed. This is the "normal" wavelength dispersion effect. Therefore, he argued, a compound wave form should disintegrate after traveling a while. This is true for a wave group that is composed of small-amplitude sinusoidal waves.

However, as Russell learned from his experiments, tall waves are subject to a new phenomenon: *amplitude* dispersion. That is, the higher the wave, the faster it travels. In addition, a tall wave is subject to the usual *wavelength* dispersion: the longer the wavelength, the faster the speed of the wave. Stokes managed to exactly balance these two types of dispersion so as to ensure a stable profile. But complex things are rarely perfect: in later work, Stokes waves were shown to be unstable after they have traveled a sufficient distance.

Stokes's theory was the first step toward describing waves whose heights are small but not infinitesimal, compared with their wavelengths. Real ocean waves are not pure sinusoids; they are much closer in profile to a Stokes wave. His theory also laid the foundation for the study of so-called solitons, waves with a large single peak that propagates without change of shape. As we will see, solitons have become very trendy because they appear as pulses in light, in plasmas, in sound, and possibly as rogue waves in the ocean. They represent an extreme form of a finite wave, and a lot of recent research is devoted toward understanding their properties.

Inching toward Theories of Tall Waves

Strictly speaking, Stokes was not the first to describe tall waves mathematically. That distinction properly belongs to Franz Joseph von Gerstner, a Bohemian physicist and railroad engineer, who published his results in 1803, long before Stokes. Gerstner adopted a form of the wave equation invented by Lagrange, which is easier to solve in some circumstances than Euler's wave equation. Gerstner was able to obtain "exact" solutions, which are compact formulas without the string of correction terms that Stokes employed.

Gerstner's solutions are waves that can travel without a change of shape in deep water. They are extreme examples of Stokes's shallow-water trochoidal waves. Their crests are sharp pointed cusps, and their troughs are very flat. And yet the orbits of blobs under the surface are circles, just like the orbits of very weak Airy waves.

A Gerstner wave with a long wavelength begins to look a lot like the solitary wave that Russell reported, a single peak traveling swiftly with a finite height. However, Gerstner's result remained buried in the literature until William J. M. Rankine rediscovered it in 1863. Even then, it was dismissed as having limited applicability to real waves because rather peculiar forces would be needed to set the wave in motion.

Diederik Korteweg and Gustav de Vries, two Dutch mathematicians, made the next important advance in the study of waves with finite heights. Korteweg, a professor of mathematics and physics at the University of Amsterdam, had a long-standing interest in wave motion, beginning with his doctoral thesis. In it he solved the problem of how a beating heart creates waves in elastic arteries.

In 1895 Korteweg and his student de Vries discovered a form of Euler's wave equation that could be solved exactly and that yielded a family of shallow-water waves with finite heights. They are trochoids with even sharper crests and broader troughs than Gerstner's waves. These Korteweg–de Vries (KdV) waves propagate without change of shape, and the long-wavelength versions resemble Russell's solitary wave closely. However, they were considered exotic and were rediscovered only after solitary waves became interesting to physicists.

Solitary Waves Abound

Solitary waves began to show up in a variety of physical situations during the 1960s, and the KdV waves proved useful in modeling them. For example, physicists working at Princeton University discovered solitary waves in the plasmas they were experimenting with. Plasma, you may recall, is a kind of gas composed of free electrons and electrically charged ions. It results from heating a substance to very high temperature.

In 1965 Norman Zabusky and Martin Kruskal, two plasma physicists, carried out pioneering numerical simulations in order to interpret their plasma waves. They discovered that KdV solutions matched the experimental solitary waves nicely and renamed the waves "solitons," a species we encountered above and will encounter again. Moreover, they found that KdV waves do not change shape when they collide; they merely exchange positions. As we will see, this result conflicts with observations of ocean waves in storms.

Particles Are Waves and Vice Versa

Physicists have continued to search for a mathematical theory that accurately predicts the behavior of ocean waves, sometimes reaching into very disparate fields. Recently, they have successfully modified a wave equation that is well known in atomic and nuclear physics. These developments have proven useful in understanding the formation of rogue waves, so I will take a moment now to describe them in a general way.

In 1927, Joseph Davisson and Lester Germer demonstrated that electrons (usually thought of as particles) could be made to produce interference and diffraction patterns, just like waves. The idea that a particle could behave like a wave, or that an extended object like a wave could represent a localized particle, was revolutionary and called for an explanation.

In response, Erwin Schrödinger, an Austrian physicist, introduced his now-famous wave equation for atomic particles that same year. The equation governs the evolution of a wave that represents the probability that a particle is located near some chosen point in space. In the strange world of quantum physics, particle positions are not determined, even conceptually, until they are measured. However, the Schrödinger wave yields the probability of several independent measurements of position.

Schrödinger used his probability wave equation to predict the energy levels of the hydrogen atom, a theoretical advance of the highest importance. He went on to predict the energy levels of a pair of rotating atoms and the splitting of atomic energy levels in a strong electric field. These results matched experiments closely and established the Schrödinger equation as the basic tool in quantum physics. In recognition for this achievement, Schrödinger shared the Nobel Prize in physics with Paul Dirac in 1933.

Then in the 1960s physicists were struck by a possible analogy between a soliton in water waves (that is, a localized particle-like structure in a wave system) and an electron (that is, a single particle whose position is determined by a probability wave). They tinkered with the Schrödinger wave equation, added some new factors, and by 1979 applied this nonlinear version to model solitons, not only in water waves but also in light waves and plasma waves.

Meanwhile, Vladimir E. Zakharov, a world-class Soviet theoretician, derived another nonlinear wave equation that was suitable for plasma. In 1968 he demonstrated that his equation also governs the envelope of a group of steep water waves and that the envelope can shrink to form a set of steep narrow solitons.

So at this point, we had two parallel theories for the origin of solitons: the Schrödinger equation and the Zakharov equation. Which was closer to the truth? In 1999 Lev Shemer and his colleagues at Tel Aviv University decided to put them to an experimental test. Both theories performed quite well, but the Zakharov model was somewhat better. We will return to these models of solitons in the chapter on rogue waves.

Storm Waves Exchange Energy

Waves with very small heights relative to their wavelengths can collide and just pass through each other with no exchange of energy. (We saw this on the beach when two pebbles were dropped into a pool.) In effect, weak waves preserve their identities in a collision and separate without damage.

But tall waves (with heights a few percent of their wavelengths) are different. To begin with, they are no longer sinusoidal in shape. As the Webers discovered with their wave tank, tall waves have sharper cusps and broader troughs than sine waves. Moreover, when tall waves overlap, they can exchange energy and change shape. More importantly, they can spawn new waves that have quite different wavelengths. These effects are examples of nonlinear behavior, and they greatly complicate the task of predicting the steady-state spectrum of ocean waves.

In 1960 Owen Phillips, whom we first met in chapter 3, took the first step in analyzing how tall waves exchange energy. Phillips studied the collision of two deep-water wave trains of arbitrary wavelengths and directions. He set up an equation to describe the time history of the sine and cosine waves that represent the instantaneous shapes of the interacting waves. He found an approximate solution, such that weak secondary waves with new wavelengths are generated in the collision. These secondary waves could not grow at the expense of the primary waves because their speeds were mismatched.

However, *tertiary* waves with a different set of wavelengths were also spawned in the collision. And they *could* interact with the primary waves, growing indefinitely at a constant rate. As a result, the original wave trains decayed into these "sidebands." Perhaps it is not surprising that a collision of two trains should result in a train wreck! But Phillips's result was quite unexpected, an indication that tall waves interact nonlinearly. That is, they generate new waves with different wavelengths, and these new waves can drain energy from their parent waves.

In the ocean, where deep-water waves of all wavelengths could interact, Phillips could envision the continuous decay of waves by the growth of their sidebands. And then the process would repeat: the sidebands would develop sidebands and decay as well. Would the wave system settle to an equilibrium state? If so, what would it look like? This was a question that only new observations could answer.

A Lab Experiment

Seven years after Phillips published his theory of colliding wave trains, a similar effect was observed in a laboratory experiment. Thomas Brooke Benjamin and his student James E. Feir were generating deepwater waves in a long tank at Cambridge University. They were using an oscillating paddle at one end of the tank to launch a train of moderately steep waves with a chosen period. The individual waves in the train would start out with well-defined profiles and an apparently constant frequency. But about halfway down the length of the tank, the wave train disintegrated. What went wrong?

The experimenters racked their brains to explain this odd result. Eventually, they traced the cause of the collapse to a slight imperfection of the paddle. It had introduced secondary frequencies in the wave train. These secondary waves had slightly larger and slightly smaller frequencies compared with the primary wave. In effect, the initial wave train was accidentally being frequency-modulated. The experimenters were able to show how energy was transferred from the primary wave to these sidebands and how they grew in amplitude explosively. This Benjamin-Feir instability was an example of wave nonlinearity: the possibility of energy transfers among finite waves. We will see that this instability has been invoked to explain extreme waves, the notorious rogues.

Brooke Benjamin was one of those rare scientists who combine mathematical ability with deep physical intuition. He contributed to the solution of many difficult problems in hydrodynamics, such as the supersonic collapse of air bubbles in water, the similarities between sea breezes and avalanches, and the mechanism through which vortices on aircraft wings could turn into shock waves.

A story is told that he was invited to dinner at Prime Minister Margaret Thatcher's apartments at 10 Downing Street in London. While mingling with the guests, he met the prime minister's husband, Denis, who asked who he was. He replied that he was a professor at Oxford University. "I should keep that quiet around here," Thatcher murmured. Benjamin recalled that Margaret had been the dean of an Oxford college and had had some serious clashes with the equally strong-minded Oxford faculty.

Hasselmann's Wave Quadruplets

Phillips's analysis of two colliding wave trains was correct as far as it went, but it was not the whole story. In 1962 Klaus Hasselmann, a young researcher at the University of Hamburg, completed a doctoral thesis in which he showed how tall waves in the open ocean could exchange energy in a collision, but only under certain subtle restrictions. Moreover, he claimed that these exchanges could be just as important in changing the relative strengths of waves as the wind's energy input.

As we will see, Hasselmann has played a prominent role in physical oceanography for the past 40 years. He debuted as a theorist in the 1960s. In the 1970s he organized a major observational campaign to measure wave spectra at sea, which I describe in the next chapter. In part, this campaign was intended to test Hasselmann's theory of wave interactions. Then in the 1980s he helped to develop a practical wave forecasting procedure, based on observations by radar-equipped satellites.

In his seminal 1962 paper Hasselmann predicted that tall gravity waves moving in independent directions could exchange energy only in groups of four (a "quadruplet") and only if the members of the quadruplet were related by direction, wavelength, and frequency. So, for example, the sum of the frequencies of one pair must equal the sum of frequencies of the other pair. This type of interaction was not merely academic, Hasselmann claimed. In fact, the transfer of energy among waves could be a central factor in the evolution of a wind-driven sea.

Hasselmann showed how to calculate the change in energy of a chosen wavelength resulting from interactions with all other wavelengths. Hasselmann's math was elegant, but his formal solution was so complicated that it could not be calculated in a reasonable time with the computers that were available in the 1960s. Therefore, his solution was not suited for real-time forecasting. Over the next two decades, Hasselmann and his collaborators continued to search for approximate solutions that could be calculated quickly. He succeeded eventually.

Enough of theory for a while. Let's move on to observations of real waves in a real ocean.

CHAPTER FIVE

Observations at Sea

The Postwar Boom

If you were the captain of a ship crossing the North Atlantic during World War II, you depended on your experience of the sea to survive. In winter the gales were ferocious, and the waves could easily top 10 m in height. In sub-zero temperatures, the seas breaking over the deck would freeze instantly and coat your vessel with ice. Your ship pitched and rolled mercilessly. Capsizing was always a possibility.

As the wind kept rising, you might ask yourself how high the waves could grow. Could your ship survive? To reassure yourself, you might consult the Beaufort wind force scale. This chart relates the average wind speed to the average wave height (table 5.1). It dates back to the 1830s, when the Admiralty, headed by Rear Admiral Sir Francis Beaufort, made the scale standard equipment in the British Navy. It was first officially used during Darwin's voyage on HMS *Beagle*. The scale originally had 12 steps, from calm to hurricane (0–71 knots, or 0–37 m/s); it was extended in 1946 to 17 steps for exceptionally violent hurricanes such as the typhoons in the northwestern Pacific. (A knot equals 1.85 km/h, or 1.15 mi/h.)

The scale served mariners well then and remains a valuable aid even today. But the wave heights listed in the chart usually represent visual estimates, not objective measurements. And as every seadog knows, local conditions can generate occasional waves significantly more brutal than the chart suggests. To improve estimates of wave heights and to be able to predict them in advance, oceanographers knew that better observations were needed.

The Glitter Experiment

Walter Munk and his colleague Charles Cox took the first early steps to measure the roughness of the sea. In 1951 they devised an experiment to relate

Table 5.1.   Beaufort wind force scale

Force	Wind (knots)	WMO classification	Water appearance
0	Less than 1	Calm	Sea surface smooth and mirror-like
1	1–3	Light air	Scaly ripples; no foam crests
2	4–6	Light breeze	Small wavelets; crests glassy; no breaking
3	7–10	Gentle breeze	Large wavelets; crests begin to break; scattered whitecaps
4	11–16	Moderate breeze	Small waves, 1–4 ft, becoming longer; numerous whitecaps
5	17–21	Fresh breeze	Moderate waves, 4–8 ft, taking longer form; many whitecaps; some spray
6	22–27	Strong breeze	Larger waves, 8–13 ft; whitecaps common; more spray
7	28–33	Near gale	Sea heaps up; waves 13–19 ft; white foam streaks off breakers
8	34–40	Gale	Moderately high (18–25 ft) waves of greater length; edges of crests begin to break into spindrift; foam blown in streaks
9	41–47	Strong gale	High waves (23–32 ft); sea begins to roll; dense streaks of foam; spray may reduce visibility
10	48–55	Storm	Very high waves (29–41 ft) with overhanging crests; sea white with densely blown foam; heavy rolling; lowered visibility
11	56–63	Violent storm	Exceptionally high (37–52 ft) waves; foam patches cover sea; visibility more reduced
12	64+	Hurricane	Air filled with foam; waves over 45 ft; sea completely white with driving spray; visibility greatly reduced

wind speed to wave height, or more precisely, to wave *slope*, which is the ratio of wave height to wavelength.

Munk had been thinking about the generation of ocean waves by the wind. He guessed that the roughness of the sea was critical to the generation of waves. The rougher the sea, the easier the wind could gain a purchase on the water. In nontechnical terms, the rougher the sea, the greater the drag of the wind on the water. And the greater the drag, the higher the waves would grow and the rougher the sea would become. It was a feedback mechanism for growing waves. (This was six years before Phillips or Miles had published their theories.)

How could one measure the roughness of the sea? Around 1950, Munk realized that the glitter of the sunlit sea, as seen from a great height, contained information on the average slope of wave crests. That could be a measure of the roughness of the sea. He reasoned as follows:

If the sea was perfectly flat, an aviator would see a single bright spot on the surface at the unique point for mirror-like reflection. That is, the geometry of the rays from the sun and to the aviator would limit the reflection to one point (the "specular" point) on the sea. If the sea was covered with randomly oriented wave crests, however, the aviator would see thousands of glinting points. At each point a crest would be correctly oriented to reflect sunlight to the observer. The farther the reflecting point is from the specular point, the greater must be its slope. Therefore, the width of the glitter pattern is a measure of the maximum slope of the waves.

Munk was not the first to think of this method. It turns out that a certain John Spooner described such observations in 1822. But Munk had the advantage of improved technology.

He borrowed a B-17 bomber from the U.S. Air Force and equipped it with four cameras. Two of these pointed straight down; the other two pointed at an angle of 30 degrees, so as to extend the field of view. Then he had the plane photograph the glitter from a height of 2,000 feet under varying wind conditions. The photographs were taken over the Alenuihaha Channel between the islands of Hawaii and Maui. "I remember Chip Cox in the transparent bubble of the B-17, leaning forward to select a site and triggering the camera with his bare toes," Munk said. To correlate the wave slope with the wind speed, Munk chartered a schooner to measure the wind speed and direction at two heights above the sea.

The photographic images had to be corrected for a background of diffusely scattered sunlight. After a tedious analysis, Munk and Cox could determine the mean square wave slope as a function of wind speed. The result was clear: the wave slope increases in direct proportion to the wind speed and reaches a maximum (equal to the square of the tangent of 16 degrees, or 0.082) at a wind speed of 14 m/s.

As we will see, this basic technique has been extended to interpret reflected microwaves (radar) from aircraft and from satellites. Moreover, once the relationship between sea roughness and wind speed is known, it becomes possible to derive the wind speed over a wide area from satellite measurements of sea roughness.

The Slopes of Storm Waves

These days, with our satellites, radar systems, and sophisticated meteorological instruments, we tend to forget how difficult it was in the 1940s and 1950s to acquire real measurements of storm waves. Most mariners were too occupied with survival in a storm to spend time observing the waves. And yet without objective measurements it would be impossible to devise a practical wave forecasting scheme.

Measurements of high waves were not entirely lacking, however. Beginning in February 1953, the British Weather Service stationed the weather ship *Explorer* at two locations in the North Atlantic Ocean. One site was in deep water, the other over the shallower continental shelf. The mission of the ship was mainly to gather wind and cloud data for weather forecasts, but the ship was also equipped with wave height recorders. This simple device measured the varying heights of the sea relative to the ship's pitching hull.

J. Darbyshire, a scientist at the National Institute of Oceanography, recognized the potential value of these records. In 1955 he decided to examine two years of wave height data. Using Fourier analysis, he could extract each sine wave's amplitude and period from the data and then compute its slope (height divided by wavelength). It must have been a tedious job, using mechanical desk calculators.

Darbyshire focused on how the slopes of waves varied in a steady wind as the fetch—the distance the wind has blown over open water—increased. He learned that for relatively short fetches, the slope of the highest waves increased with increasing wind speed and *decreased* with increasing fetch, contrary to what one might guess. The decrease continued until, after a fetch of about 100 miles, the slopes reached a steady equilibrium value.

He discovered a second effect as well: at fetches larger than 100 miles, the slopes of the highest waves were greater at higher wind speeds on the deep ocean, as one might expect. But in shallower water the equilibrium slope *decreased* at higher wind speeds. Darbyshire pointed to the different patterns of wave growth but had no ready explanation for the differences. Nevertheless, he had obtained the first reliable description of the evolution of waves and how it depends on fetch and wind speed.

Lionel Moskowitz and Fully Developed Seas

During the next decade the British Weather Service expanded its fleet to four weather ships, each equipped with a wave recorder. In 1963 Lionel Moskowitz, a relatively obscure analyst at the U.S. Naval Oceanographic Office, realized that the wave records of these ships contained a treasure trove of information if it could be analyzed properly. He wanted to find occasions when the sea had come to equilibrium with the wind—that is, when the wind had blown steadily over a sufficiently large distance (fetch) and for a sufficiently long time (duration) for the wave spectrum to reach a steady state. Such a sea he defined as "fully developed," which, he thought, would be the one of most interest to mariners.

Moskowitz began by examining weather maps for the North Atlantic produced every 6 hours by the U.S Weather Bureau. He selected intervals in 1955 through 1960 in which the wind near a weather ship was steady in speed and direction. He found 460 examples, with wind speeds ranging from 20 to 40 knots (about 23 to 46 mph) in steps of 5 knots. For each case he estimated the fetch and duration of the wind. Using these parameters, he could identify wave records that corresponded to fully developed seas. Each wave record was usually rather short, lasting only 7 to 15 minutes. But he proceeded to digitize each time series and to Fourier-analyze them in order to obtain their energy spectra.

Moskowitz was a careful worker. He was rightly concerned that these short records might not be representative samples of the true distribution of wave heights at a fixed wind speed. So he applied a statistical test and eliminated records that failed the test. Only then did he combine the spectra to obtain an average energy spectrum for each wind speed.

Each average spectrum was a peaked curve of energy versus frequency, similar to the curve in figure 4.1. Moskowitz learned that the peak of the energy spectrum rises and shifts to lower frequencies (longer wavelengths) as the wind speed increases. That is, the bulk of the energy is increasingly carried in long wavelengths. He was able to calculate the so-called *significant wave height* at each wind speed, which Walter Munk and Harald Sverdrup (the director of the Scripps Institution) had earlier defined as the average height of the one-third highest waves. He determined that the significant wave height increases as the square of the wind speed. So, for example, the significant wave height at wind speeds of 20 and 40 knots are 7.3 and 29.1 feet (2.2 and 8.9 m), respectively. This was the kind of practical information a sea captain could use.

Self-Similar Spectra

But this was only the beginning for Lionel Moskowitz. Willard Pierson, a professor of oceanography and meteorology at New York University, learned of Moskowitz's work and immediately got in touch with him. Pierson was familiar with an idea that Soviet theorist S. A. Kitaigorodskii had proposed for waves in a fully developed sea. Taking a clue from turbulence theory, Kitaigorodskii proposed that the spectra of waves at different wind speeds might be "self-similar." That is, when plotted in certain nondimensional variables, the wave energy spectra would all have the same shape and could be folded into one universal shape. This would imply that other factors, like fetch of the wind, would not affect the wave energy spectra.

Pierson and Moskowitz collaborated in testing this idea mathematically using Moskowitz's data. Despite some discrepancies and some necessary adjustments, the data fit the Soviet scientist's hypothesis reasonably well. All of Moskowitz's spectra collapsed into a single shape. That result seemed to lead to two conclusions: that fully developed seas are fairly common in the North Atlantic and that their energy spectra are determined by only one parameter, the wind speed. The significant heights and periods of waves both increase in proportion to the wind speed, according to the Pierson-Moskowitz spectrum.

These results were so invaluable for forecasting conditions in massive storms that the Pierson-Moskowitz spectrum enjoyed a fine reputation for a few years. However, it was unsatisfactory in at least two respects. First, it predicted that a wave with the frequency of the peak of the spectrum could propagate 17% faster than the measured wind, a puzzling situation, to say the least. Secondly, there was no independent evidence that Moskowitz's data corresponded to "fully developed" seas. In fact, some experts continued to doubt whether such conditions ever exist in nature. It seemed more likely that a spectrum would always depend on the fetch and duration of the wind, as well as its speed. To settle the issue, oceanographers began to think about acquiring more detailed data in a large-scale experiment at sea. They had a good example to guide them.

The International Geophysical Year

On April 5, 1950, a group of geophysicists held a meeting in Silver Spring, Maryland, to discuss the work of Professor Sydney Chapman, who was visiting from England. Chapman had advanced some interesting ideas about the sun's effect on the earth's magnetic field. After the formal meeting, the talk shifted to other areas of geophysics. The scientists realized that technology had advanced considerably since World War II. New tools were available to explore the atmosphere, the earth, and the oceans. Radar that could "see" great distances, electronic sensors that could provide scads of raw data, computers that could crunch these large amounts of data, and rockets that could reach up into space were now at hand. The idea emerged for a major campaign to study a vast array of problems in geophysics and meteorology.

The time was ripe for such an enterprise. During the war many nations came to realize how their security and economies depended on a detailed knowledge of the planet on which they lived. Government agencies were willing to help fund a large cooperative program. Over the next seven years, many nations and institutions developed ambitious plans under the organizational umbrella of the United Nations to carry out an extensive program of observations. Research was to be done in such specialized fields as aeronomy (the study of the upper atmosphere), geodesy (the measurement of the size and shape of the earth), geomagnetism, ocean circulation, and bathymetry of oceans. An 18-month period, July 1957 through December 1958, was designated as the International Geophysical Year (IGY). In the end, 67 nations participated in research.

The IGY was hugely productive. The concept of continental drift was confirmed. The discovery of deep submarine trenches and an immense submarine mountain chain winding 64,000 km around the earth led eventually to the development of the theory of tectonic plates. The Van Allen radiation belts around the earth were discovered, ocean currents were mapped, Antarctica was explored, the geomagnetic field was studied, the first satellites were launched into orbit (first by the Soviet Union, much to the chagrin of the United States), and winds in the upper atmosphere were measured. Finally, detailed data about our world were being collected and analyzed in ways not previously feasible.

An Expedition for Wave Physicists: Hasselmann and JONSWAP

The scientists of the IGY explored a vast range of topics, but the generation and dynamics of ocean waves were not among them. Nonetheless, the IGY spirit and accomplishments inspired physicists and oceanographers who were studying waves to take advantage of the technology and new data analysis methods to embark on a similar kind of cooperative campaign.

Klaus Hasselmann was among the first to recognize the need for better data. In 1966, at the age of 35, his reputation as a theorist had won him the chair of a small theoretical group at the University of Hamburg. Four years earlier he had published his theory on the energy transfer among ocean wave quadruplets; now he was eager to test it.

Hasselmann became interested in science as a teenager. He was fascinated by the crystal detector, a crude radio, which allowed him to hear beautiful music without buying a radio. To find out how the crystal detector worked, he went off to the local library, where he read widely about electricity and physics. Like any teenager, he was also interested in airplanes and rockets. He built electrical motors and other gadgets, occasionally short-circuiting the wires in his home but learning from his mistakes as much as from his successes.

He chose physics as a career, studied in Hamburg, and in 1957 wrote a doctoral dissertation on turbulence and a type of elastic wave that propagates at the boundary between two layers. He gradually drifted into oceanography, drawn initially by the problem of wave resistance to ships. That led to his major theoretical breakthrough, a mathematical description of the resonant energy transfer among ocean waves. In 1961 he presented his work at a conference in the United States and as a result was invited to join the Scripps Institution by Walter Munk. He spent three fruitful years there and then returned to Hamburg as an assistant professor.

Hasselmann had been impressed by Munk's multi-ship experiment to investigate the decay of swells across the Pacific. He began to think about a similar program to study the evolution of ocean waves. In 1966, at an international conference in Bern, Switzerland, Hasselmann corralled an informal group of colleagues to discuss undertaking a massive experiment at sea. It would be designed to gather the most complete set of measurements on wind-driven waves. The idea was to measure the change in the wave energy spectrum as it develops from short waves near the shore, to longer waves further offshore, and to the higher and longer waves in a fully developed sea, if such a state really exists.

After the conference the group scattered to their home institutions to promote the program. It was the ideal moment to propose such an experiment. As a result of naval experience in World War II and the success of the IGY, many governments were beginning to realize the importance of oceanography. Thus came the formation of the Joint North Sea Wave Project (JONSWAP), a collaboration among dozens of scientists from Britain, the United States, Germany, and Holland. Funding was provided by the governments and private sources as well as by NATO. The scientists in the group chose a site off the island of Sylt, a famous nudist resort on the border between Germany and Denmark, as the base of their endeavor.

Hasselmann reluctantly agreed to act as the coordinator of the venture, despite his lack of experimental credentials. He turned out to be an excellent organizer. He was responsible for acquiring the ships and personnel needed and deploying them at the site. In addition, he designed a detailed daily schedule for the observations that the ships would make over several months at sea. The ships would be connected by radio telephone, allowing close coordination of the observations. The participants were ready to launch the experiment by the summer of 1968, but it had to be postponed to the summer of 1969 because of naval exercises in the North Sea.

Thirteen stations were laid out on a line that extended 160 km northwest of Sylt. In an east (offshore) wind the array of instruments could measure waves under conditions of limited fetch (recall that the fetch is the distance over which the wind blows with constant speed and direction). In a westerly wind the array could measure the dissipation of swells as they rode inshore. Detailed measurements of winds, waves, and currents could be made over a period of three months. Many kinds of instruments were employed. These included accelerometers (to measure rapid changes in water movement), pressure-sensitive buoys (to measure wave heights), pitch-and-roll buoys (to determine the directions of waves), wave-followers (equipped with a chain of pressure sensors above and below the surface), and anemometers (to measure the wind speed at a range of heights).

The observing program went off brilliantly. Over 50 million data points were collected. Because of the primitive computing facilities available at the time, this mountain of data was difficult to analyze; much of the data had to be copied to tapes or to punched cards by hand. Not until 1971 were the final results in hand; they were published in 1973, seven years after the conception of the experiment. Hasselmann recalled the experience fondly in an interview made in 2006. "It was great fun," he said, "and we made many long-lasting friendships. We worked hard and relaxed at parties. Everyone was included: the crews, the telephone operators and the technicians."

Figure 5.1 shows how the wave energy spectra varied systematically as a steady wind (10 m/s) blew over longer and longer fetches. (Remember that frequency is the inverse of period—so, for example, a frequency of 0.1 Hz corresponds to a period of 10 seconds—and that long periods correspond to long wavelengths.) In the figure, the curves are labeled with the station number, with station 11 located furthest downwind or at the greatest fetch; station 5 had the smallest fetch. The message these spectra convey is that fetch matters: a steady offshore wind that blows for a long time generates the highest and longest waves at the greatest fetch. Station 5 never saw the energetic waves or long periods that station 11 did, no matter how long the wind blew.

A fully developed sea, in which the wind and the waves are in equilibrium, was never observed at any fetch or at any wind speed during the JONSWAP experiment. However, it was possible that even the longest fetch in the experiment (at station 11) was still too short to develop a fully developed sea. So the existence of a fully developed sea was left undecided.

Like the Pierson-Moskowitz spectra, the JONSWAP spectra were *self-similar*: they had the same basic shape despite differences in size. In fact, all these spectra could be represented by one master formula with adjustable parameters for fetch and wind speed. This JONSWAP master spectrum resembled the Pierson-Moskowitz spectrum but had a much higher peak for the same wind speed. The JONSWAP spectrum became the preferred tool for forecasters. By estimating the fetch and wind speed in a hurricane, a forecaster could use the master formula to estimate the heights of the tallest waves. A great boon to the mariner!

Another major conclusion drawn from the experiment concerns the transfer of energy among waves of different wavelengths, the process that Hasselmann had predicted in 1962. He predicted that tall waves moving independently could exchange energy only in groups of four, and only if certain relationships held. This process turned out to be the dominant one for wave growth and accounted for the continuing evolution of the wave spectra.

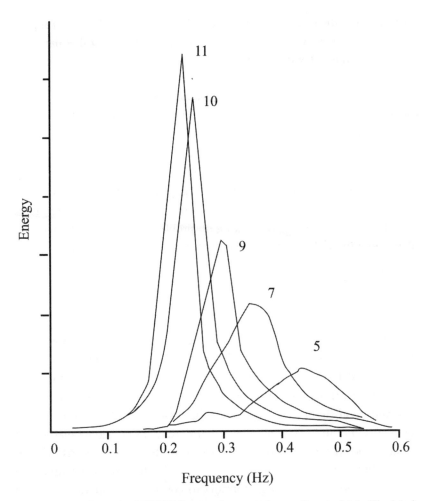

Fig. 5.1 Spectra from the JONSWAP campaign sorted according to fetch. The labels indicate station positions from shore, so, for example, station 11 has the greatest fetch for east winds (120 km), and station 5 has the least. (Drawn after T. P. Barnett, Offshore Technology Conference, Houston, Texas, May 3–5, 1972, fig. 3.)

The picture that emerged for the flow of energy was as follows. Wind energy is deposited in the sea at short wavelengths, possibly by the Miles mechanism. Then, precisely as Hasselmann had predicted, wave energy is redistributed to longer and longer waves. As a result, waves with frequencies near the peak of the wave energy spectrum can (surprisingly) propagate faster than the wind, their excess energy deriving from wave-to-wave energy transfers. The long waves ultimately dissipate by some process like turbulence or whitecapping. How waves dissipate was, in fact, the most uncertain question still remaining.

The successful JONSWAP campaign gave Hasselmann's career a tremendous boost. His theory of wave-to-wave energy transfer was not only confirmed but turned out to be the dominant process in wave growth after the initial phase of wave generation. Moreover, he had demonstrated an ability to manage a complex science project that involved both observation and analysis. He went on to become a full professor of theoretical geophysics at the University of Hamburg and, later, the director of the Max Planck Institute of Meteorology. Recently, he wrote a fascinating book, part history of oceanography and part biography, based on a long weekend of interviews with Walter Munk, his old mentor. He is presently working on problems of climate change, having developed the well-respected Hasselmann model of climate variability.

The Spreading of Storm Waves and the SWADE Project

In the early 1950s Willard Pierson and his co-workers at the U.S. Hydrographic Office were developing the first practical methods of forecasting wave spectra, basing their work on scattered data obtained from weather ships. While analyzing the data, they discovered that waves in a strong wind do not all travel exactly in the wind's direction. They recognized that this spreading effect could be an important process in the dissipation of storm waves and therefore essential to forecasters. They tried to describe the effect mathematically but found that the spectrum of waves spreading in directions different from the wind depends in a complicated way on fetch, wind speed, and duration. So they settled for a rough formula to cover the broad variation they found.

Better data from the JONSWAP campaign led Klaus Hasselmann and colleagues to a clearer description of the energy spectrum of these spreading waves. They found that waves with frequencies within 10% of the spectrum peak's frequency are narrowly focused and travel within 30 degrees of the wind direction. However, waves with higher and lower frequencies spread out in directions as far as 60 degrees from the wind. In other words, the most energetic waves propagate nearly downwind, while the less energetic waves spread sideways.

After Hasselmann published these results in 1980, observers continued to argue about the various factors that affect the spreading of wind-driven waves and proposed a great variety of fitting formulas. The argument was not academic because predictions of wave energy spectra depend sensitively on the spreading effect. By the late 1980s oceanographers began to talk about mounting another experiment comparable in scale to JONSWAP. It would fully explore the variation of wave energy spectra with *direction* and investigate a new topic: the evolution of the wave energy spectrum in a *changing* wind.

This was a good time to plan another large campaign. The new observational tools that had been invented since the JONSWAP campaign promised to deliver much improved data. In particular, an aircraft equipped with a radar altimeter (described in a later chapter) could provide detailed observations of the wind over a huge area. A new generation of moored meteorological buoys was also available for wind and wave measurements.

So between 1986 and 1989 a team of 50 scientists from the United States and Canada designed plans for what would be named the Surface Wave Dynamics Experiment (SWADE). Robert A. Weller (Woods Hole Oceanic Institution) and Mark A. Donelan (Canadian National Water Research Institute) were the main organizers. The program would employ a great variety of research platforms and a huge team of scientists, technicians, and engineers to acquire and analyze the data. Funds for this effort would be provided by the U.S. Office of Naval Research, NASA, and the National Oceanic and Atmospheric Administration (NOAA), as well as by several European governmental agencies.

The campaign was planned to begin in October 1990 and last until March 1991. During the summer of 1990 the equipment was assembled off the eastern coast of the United States. An array of over a hundred instrumented buoys was moored between 50 km and 500 km off the coast, from North Carolina to Rhode Island. For the first time some of these instruments would be able to measure the direction of the wave field as well as its strength. Simultaneous observations of the surface winds would be carried out by radar-equipped aircraft in a series of two-week periods of intensive operations.

In late October, just as the program was to begin, a massive storm wiped out many of the buoys. Imagine the disappointment and frustration of the scientists! The program was delayed until the buoys could be replaced or substitutions could be found.

Except for this setback, the scientists were lucky. They were able to gather data during the three distinct types of storm conditions they most wanted.

First, they wanted a severe cold front with rapid shifts of wind direction, for studies of the response of the wave spectrum. Second, they wanted a steady off-shore wind, for short-fetch wave growth studies. Finally they hoped for long fetches from the northeast so that they could study swell propagation. All three conditions were encountered during the program. And as a bonus, the Gulf Stream migrated north into the SWADE study region, an event that offered unique information on the interaction of waves and currents.

A real-time forecasting program was carried out as an essential part of the campaign. Wind data gathered from the buoys was immediately fed into an advanced prediction code (3GWAM, discussed later) that was used to forecast wave heights and directions. These forecasts were later compared with the nearly simultaneous observations of the sea surface, a start on what eventually became "validation," as we will see in a later chapter.

In the following sections, I describe two experiments among the many executed during the SWADE campaign. In the first, William M. Drennan, a professor at the University of Miami, and his team were interested in how a preexisting swell might interact with a wind-driven sea. In the second experiment, Iohannis K. Tsanis and F. Brisette were concerned with the changes in wave spectra as the wind changed direction.

Growth of Waves in a Cross-Sea

For their study of preexisting swells, Drennan's team had counted on using the SPAR, an advanced spar buoy, as part of their equipment. When the October storm wiped out the SPAR, the team had to improvise. They decided to mount their sensors on a small waterline area twin hull (SWATH) vessel.

They chose their ship with care. The *Creed* was a small catamaran, only 20 m long, that slid through the water on two narrow pontoons with a minimal disturbance of the water. The researchers mounted their array of six underwater capacitance gauges from a vertical spar at the end of a long boom that extended from the starboard side of the ship. The array allowed them to measure the variation of wave spectra with time and direction, while an anemometer collected wind data from a mast.

With the ship cruising slowly into the wind, the team obtained 29 hours of useful data. They were limited by the material strength of their sensors to relatively quiet periods in which the significant wave height was less than 2.8 m. Their results are interesting nevertheless, as figure 5.2 illustrates. This is a typical polar plot, in which directions relative to north appear as

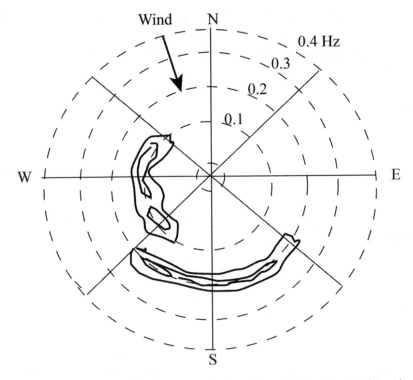

Fig. 5.2 Storm waves spread in directions far from the wind direction. In this polar plot of wave spectra, the arrow designates the wind direction. The contours indicate levels of wave energy, with wave frequencies indicated on the circles; the smallest circle corresponds to 0.1 Hz, the largest to 0.4 Hz. The broad, irregular arc at the bottom of the plot contains the wind-driven waves, which have frequencies around 0.2 Hz. The C-shaped arc on the west is a persistent swell, a remnant of winds of the preceding day. (Drawn after W. M. Drennan et al., *Journal of Atmospheric and Oceanic Technology* 11 [1994]: 1109, and used with permission of the American Meteorological Society.)

radii and frequencies from 0.1 to 0.4 Hz appear as dashed circles. The wind direction is shown by the arrow. The contours are lines of constant wave energy.

The wind on this day was blowing from the northwest, and the low-frequency waves (0.2–0.25 Hz) ran in a broad arc away from the wind. On the two preceding days the wind had blown from the east (the right side of the diagram), and a swell of lower-frequency waves (0.1–0.2 Hz) was still running westward in a 60-degree arc. So on the day represented by this diagram, waves were running in nearly every possible direction away from the prevailing wind. This was an example of a dangerous "cross-sea," a mixture of swell and storm waves with a significant height, nearly 3 m.

The researchers measured the effect of the relative directions of wind and swell on the growth rate of the wind-driven waves. As expected, the presence of the swell increased the roughness of the sea surface and therefore the rate at which waves grew taller. The biggest effect occurred when the swell and sea were running in nearly opposite directions. But the big surprise was the size of the effect: the growth rate could be increased by a factor of 3. The experiment showed that swells had to be included in any prediction of wind-driven wave heights.

Waves in a Turning Wind

Another interesting result from the SWADE campaign concerned the changes in the wave spectrum in a turning wind. I. K. Tsanis and F. Brisette, two Canadian scientists, addressed this question with SWADE data. They learned that in a slowly turning wind, the whole spectrum adjusts smoothly. In more rapid turns of the wind, a portion of the wave energy spectrum decouples from the wind and decays. In even more rapidly turning wind, the entire spectrum decouples and decays while a new spectrum, consistent with the wind direction, is born. They were able to determine the rate at which the wave energy spectrum adjusts. They found good agreement with cruder results from JONSWAP and with the predictions of an advanced model.

Testing Miles's Resonance Theory, Again

You may recall John Miles's resonance theory of wave generation, outlined in chapter 3. For many years after the theory was published (1957), several experimenters attempted to test the theory at sea. They tried to measure the changes in air pressure at the water surface that are induced by the waves and act to amplify the waves. These measurements are difficult, and the results from several attempts differed by as much as a factor of 10. The experimenters agreed, however, that the rate of exponential growth of wave energy predicted by the theory appeared to be too small by at least a factor of 3.

Miles's theory had gotten a boost in 1991, when Tihomir Hristov and friends detected a pattern of wave-induced pressure changes in good agreement with the theory. They did not, however, confirm the absolute value of the growth rate, so the accuracy of the theory was still open to question. Mark Donelan was similarly skeptical. In his view and that of a group of his

Australian co-workers, Miles's theory had never been proven beyond reasonable doubt. One of these colleagues, William Plant, said that Donelan usually refuses to accept conventional wisdom. "He goes and examines it himself" (*Soundings*, Florida State University, March 2002).

Therefore, in 1999 Donelan teamed up with the Australian Ian Young of Swinburne University of Technology, Melbourne, to run an experiment—the Australian Shallow Water Experiment, or AUSWEX—to test the Miles theory. They suspected that the mismatch between theory and observation was linked to the slopes of waves. (Slope, you will recall, is defined as the ratio of amplitude to wavelength.) Basically, a large slope would enable the wind to drag on the wave efficiently. The original Miles theory took no notice of wave slope and predicted that the rate of wave growth would depend solely on wavelength and the vertical profile of wind speed.

A wave's slope increases with the wind speed. Or more accurately, the slope increases with the ratio of the wind speed to the wave speed: the larger the ratio, the larger the slope. In previous experiments this "forcing" ratio was less than 3. Donelan's team crafted an experiment in which the ratio, and therefore the slope, could be much larger.

For their purpose the team chose a shallow lake—Lake George, in New South Wales, Australia. The wind blows steadily over a long fetch on this lake. And at the downwind shore, the lake depth is less than the dominant wavelength, which is typically 30–40 cm. As a result the waves are shallow-water waves, whose speeds are depth-limited at any wind speed. Consequently, the wind-to-wave speed ratio could be quite large, and the waves could then grow to become very steep.

To measure the slope of waves, the team used a wave-follower that Donelan had developed at the University of Miami. The instrument was mounted offshore on a fixed platform. It consisted of an array of three sensitive air pressure sensors mounted on a vertical shaft positioned just above the water's surface. A computer-controlled motor raised and lowered the array rapidly so as to maintain the array at a constant height above a passing wave. The wave-follower recordings were supplemented by a complete set of measurements in the atmospheric boundary layer, on the surface of the lake, and in the water.

In the summer of 1999 the team made a series of runs in which the wind-to-wave speed ratio ranged from 5 to 8, and the wave slope was correspondingly larger than ever seen before. The experimental results showed that, as hypothesized, the exponential growth rate of a wave depends on the present slope of the wave, in addition to the factors Miles had considered. The growth rate falls to zero as the slope approaches zero. Thus, Donelan and the Australian team showed how the match between previous observations of growth rate and Miles's theory was improved by taking these new factors into account.

In addition, the experiment showed that at sufficiently high wind speeds, the airflow detaches from a crest and reattaches on the windward side of the next crest downstream. In effect, the wind skips over the troughs and leaps from crest to crest. As a result, the wave-induced air pressure changes that lie at the heart of Miles's theory are severely reduced at high wind speeds. This finding seemed to give new life to Sir Harold Jeffreys's 1924 "sheltering" theory. As described in chapter 3, Jeffreys had found that the back side of the crest shelters the front side of the crest, so the pressure on the front side is lower. As a result, the wind exerts more horizontal pressure on a crest's back side than on its front face, thus increasing the wave's height.

Whitecaps—What Causes Them?

Finally we turn to observations of whitecaps. Anyone who has watched the sea in a rising wind is familiar with the sight of the crests of waves breaking spontaneously. As the wind grows stronger, the waves get taller until their crests suddenly collapse in a froth of white foam called a whitecap. Whitecaps are now considered the most important mechanism for limiting wave growth. Therefore, oceanographers need an adequate model of whitecaps in order to predict maximum wave heights in a storm. And meteorologists need good data on whitecaps because they release aerosols like salt into the air and help to exchange heat and momentum with the air—an important mechanism relevant to studies of climate change.

Despite their importance, whitecaps are still rather mysterious. They are difficult to study at sea because they are sporadic and seemingly random. So until recently, observations were sparse and sometimes conflicting. In addition, they are difficult to model mathematically because they are essentially nonlinear dynamic events that grow exponentially after an abrupt start.

The simplest question one can ask is how much of the sea is covered with whitecaps at any given wind speed. Even this question is not easy to answer, as Duncan Ross and Vincent Cardone (both from New York University) discovered in 1974. They photographed the sea from an aircraft and measured the area covered by whitecaps at wind speeds that ranged from 10 to 20 m/s. But they found it difficult to distinguish between whitecaps that were still in progress and the white foamy streaks that persist after whitecaps have dissipated. Only after the streaks were eliminated in the analysis could they obtain a crude estimate of the whitecap coverage.

Since 1974 a score of investigators have made visual and photographic records of whitecap coverage. The latest and most detailed results were obtained in 2007 by the *Discovery*, a British research ship in the North Atlantic, and by the *Polarfront* in the Norwegian Sea. Digital cameras were used to record the percentage of area covered by whitecaps under varying wind conditions. To date, more than 8,000 whitecaps (about a quarter of the total set) have been measured with an automatic extraction program. The results show that whitecapping begins at a wind speed of 5 m/s and increases to 8% of the area at 25 m/s.

Gross results like this are useful, but what is really needed for modeling whitecaps is some criterion for when and why a wave becomes a whitecap. In 1986 Leo Holthuijsen and T. Herbers (Delft University of Technology, Holland) set up a simple experiment to find one. They deployed a single instrumented buoy in the North Sea that recorded the elevation of the sea surface and wind speed continuously. They watched the buoy with binoculars and recorded the time when a whitecap occurred near it. From their data they could determine properties such as slope, asymmetry, wavelength, and height for each wave that passed the buoy. They could also compile the numbers of breaking and nonbreaking waves. As expected, the fraction of waves that broke depended on wind speed. They found the fraction increased from 0.1 to 0.16 as the wind speed increased from 8 to 12 m/s.

Holthuijsen and Herbers failed to find a reliable physical criterion for wave breaking. None of the expected factors—the slope of the wave crest (the ratio of height to wavelength) or its asymmetry, or the wave period—was dominant. In fact, waves broke at a slope much smaller than contemporary theory predicted. But they did find one clue to the elusive criterion: the tendency for breaking waves to occur more frequently among *groups* of waves with unusually large mean heights. In fact, 70% of all breaking waves occurred within such groups.

Progress stalled until 2000, when Michael Banner (University of New South Wales) and his Australian colleagues analyzed data that had previously been obtained at three locations with very different conditions of wind fetch and speed. These locations were the Black Sea, Lake Washington (in Washington State), and the Southern Ocean. These investigators focused on Holthuijsen and Herbers's clue that waves break more frequently in groups of waves than alone. In essence, they wanted to test the idea that *energy transfer among waves in a group* was the dominant process causing wave breaking, not just high wind speed. Moreover, Banner and his associates were guided by their own new theoretical work, which predicted that the *mean slope* of the wave group (not its mean height) was the major criterion for a high probability of wave breaking.

In analyzing the data, they found that indeed the mean slope of the wave group is a better predictor of wave breaking than wind speed or wave height. Specifically, the frequency with which waves in a group break increases as the square of the mean slope of the group; including the effects of wind speed raised the probability of wave breaking only marginally.

The data thus implied that the internal dynamics of a group largely determine when waves break. Isolated waves just do not break as often. And, contrary to expectations, the speed of the wind only plays a secondary role. These results were useful clues, but they raised more questions than they answered. Much more research would be needed to explore the dynamics of wave groups and the effect they have on wave breaking.

The Dissipation of Whitecaps

As we saw above, wave breaking (whitecapping) is now regarded as the primary mechanism limiting the growth of wind-driven waves. But how much energy does a wave lose when it breaks? How does the rate of loss depend on the characteristics of the wave and the wind? Without a realistic model of wave breaking, an accurate forecast of maximum wave heights in a storm would be impossible. Oceanographers have therefore struggled to describe the process.

The past 10 years have seen a major advance, based partly on laboratory experiments and partly on a theoretical idea first proposed by Owen Phillips in 1985 that has now been validated by tests at sea.

Useful insights into the dissipation of whitecaps first came from towing tank experiments, especially those of James Duncan of the University of Maryland. In 1981 Duncan towed a submerged hydrofoil (shaped like an airplane wing) in a long water tank and measured the turbulent flows and mixing of layers in the tiny whitecaps he generated. Using very sensitive equipment, he was able to measure the rate of energy dissipation. From his work, from wave tank experiments by W. Kendall Melville (Scripps Institution of Oceanography), and from theoretical arguments by Owen Phillips, a tentative formula emerged for the rate of energy dissipation *per unit length* along the wave front: the rate of energy loss depends on the fifth power of the breaking wave's speed.

Phillips had a deep insight in 1985. At any given moment, he wrote, there will be a certain number of waves breaking within a prescribed area of ocean. Suppose you could measure the length of each whitecap along the crest of a wave as well as the wave's forward speed. Then you could plot the total of all whitecaps that have a specific speed. This would be a "whitecap distribution function," a kind of spectrum of whitecap lengths, sorted according to wave speeds. From this information, the number of breaking waves that pass a fixed point at each speed could be calculated. Moreover, if the energy dissipation rate of a whitecap is actually proportional to the fifth power of its wave's speed, one could use the whitecap distribution function to calculate the total rate of energy dissipation per unit area. This would be the missing link necessary for the practical forecasting of wave heights within the area.

Intrigued by this insight, but without the necessary technology until 1999, Melville and Peter Matusov (also at the Scripps Institution of Oceanography) finally were able to carry out an experiment off the coast of North Carolina to test Phillips's theory. They equipped a small airplane with a downward-pointing digital video camera, a laser altimeter, and a global positioning system (GPS) to continuously locate the plane relative to the shore.

They carefully measured the length and speed of each whitecap on the video frames and calculated the whitecap distribution function from this data. The results matched the predictions from Phillips's theory nicely. That means that regardless of how whitecaps evolve in space and time, they obey a universal distribution of lengths. Moreover, these results showed that the total length of whitecaps with any specific speed increases as the cube of the wind speed.

With these results in hand, Melville and Matusov could combine the rate of energy dissipation per meter of wave front with the Phillips whitecap distribution function to obtain the total rate of energy dissipation at any wind speed. This breakthrough was critical for practical methods of forecasting wave heights, as we shall see in the next chapter.

CHAPTER SIX

Forecasting and Monitoring Storm Waves

Storms at sea, and the violent waves they produce, have terrorized seamen since they first ventured onto the oceans millennia ago. Until recently, only the wisdom of ships' captains, built from many years of battling the seas and experiencing the wrath of the capricious waves, provided any help in surviving these storms. Since the 1930s, meteorologists have been able to provide increasing support in predicting the wind patterns. With constantly improving networks of weather stations, and more recently with satellites, meteorologists have been able to forecast storm winds and icing conditions several days in advance.

But until the 1960s, predictions of the effect of these winds on the ocean waves were rudimentary, and forecasts of wave heights during storms were quite unreliable. Mariners were still under threat from giant waves they could not anticipate. Forecasters vowed to do better. But as we have seen, just understanding how high a wind can raise the waves has been a daunting challenge.

Over the last 60 years oceanographers have learned, painfully at first, to forecast the heights and periods of storm waves as much as a few days in advance. In this chapter we'll see how they developed and tested their techniques, beginning in World War II. But to get a feeling for how far along they have come in their quest, we take a moment to remember one of their greatest successes—and failures: Hurricane Katrina.

Hurricane Katrina

The residents of New Orleans and of other cities along the U.S. Gulf Coast will never forget Hurricane Katrina. Katrina was born on August 23, 2005, near the Bahamas as a tropical depression and grew rapidly in intensity as it swept across Florida into the unusually warm waters of the Gulf. On August

28 Katrina became a monster category 5 storm with sustained winds of 175 mph. These hurricane-force winds extended 105 miles out from the center. A buoy 50 miles off the coast recorded waves 55 feet tall in the open sea.

At 4:00 p.m. CDT, Max Mayfield, director of the National Hurricane Center in Miami, issued a warning that a storm surge of 18 to 22 feet above normal tide level could be expected, with 28 feet in some localities. (A storm surge is a hill of water that is raised by the low atmospheric pressure in a hurricane and driven shoreward by the wind.) He cautioned that "some levees in the Greater New Orleans area could be overtopped."

Katrina had declined to a category 3 storm when she made landfall at Buras-Triumph, Louisiana, on August 29. But she still retained tremendous energy and record breadth (240 miles across) for a Gulf hurricane. Now the storm's waves rode atop a storm surge (also known as a storm tide) that the swirling winds had piled up. It was this combination of battering waves and a high storm surge that caused most of the record devastation inland.

Max Mayfield, in a March 7, 2007, interview on YouTube, recalled that he expected the storm surge to weaken after the winds of Katrina dropped to a category 3, but in fact the surge remained powerful because of the sheer size of Katrina. As bad as the winds of a category 3 hurricane are, it was that massive surge that caused the flooding of 80% of New Orleans as it raced through the narrow water channels to the city and broke through 53 levees.

Katrina caused over $75 billion of property damage and took more than 1,800 lives throughout the Gulf states. A third of the population of New Orleans left the city permanently, and reconstruction efforts were still under way more than 7 years after the disaster. The city of New Orleans will probably never fully recover from this storm.

Katrina could have caused many more fatalities if forecasters at the National Hurricane Center in Miami had not been able to predict both the path and the intensity of the storm days in advance and warn the population. Moreover, forecasts of the height of the waves that topped the storm surge proved to be vital. However, not even the forecasters anticipated the devastating power of the surge—a wall of water pushing relentlessly against puny levees and floodwalls.

New Orleans will always be under threat from hurricanes; even the most sophisticated hurricane forecasting that tries to model the complex interactions between wind, water, and land still cannot predict the possible course of a hurricane's destruction.

The Origins of Wave Forecasting

Like many other advances, the science of forecasting waves was launched during World War II. In 1942 Harald Sverdrup and Walter Munk at the Scripps Institution of Oceanography were working on antisubmarine detection for the U.S. Navy. Sverdrup was the director of Scripps, and Munk was a freshly minted Ph.D. from the California Institute of Technology and Sverdrup's former student.

In an interview in June 1986, Munk recalled that while working temporarily at the Pentagon in 1943, he had learned that practice landings for an amphibious invasion of North Africa had to be cancelled whenever the swell was higher than 7 feet. The reason was that the landing craft would turn sideways and swamp in such a swell. Munk visited the Army practice site in the Carolinas to see for himself. He returned to Washington to look up the wave statistics for North Africa and realized that high swell could be a real problem for the invasion. Many troops would drown while landing unless the swell was lower than 7 feet. Practice landings could always be cancelled at the last minute, but real landings, involving thousands of troops, boats, and supplies, were irreversible.

Apparently nobody had realized that this could be a problem. When Munk warned his superiors, he was scoffed at. He was only a junior oceanographer. Surely, they told him, "someone" was dealing with the problem. Munk would not be rebuffed. He called his mentor, Harald Sverdrup, the director of the Scripps Institution of Oceanography and told him of his fears. Sverdrup came to Washington immediately and warned the generals. Because of his scientific credentials, he was taken seriously. The two men were assigned the task of finding a solution to the problem.

So Munk and Sverdrup began to work to predict the height of the swell in North Africa. To do so, they had to examine the whole process by which storm waves grow and turn into swell, how swell travels great distances, and how offshore topography can focus wave energy from swell into high surf.

Observations had shown that as a wind blows over a lake, the wave heights and periods at the upwind shore quickly reach a steady state. If the wind blows long enough, a steady-state region expands over the whole lake, with higher waves at the downwind shore owing to the longer fetch. But if the fetch is unlimited, as may happen over the ocean, waves grow in height at the same rate everywhere. These facts helped to guide Munk and Sverdrup as they created their theory.

First, they had to learn how to estimate the wind's characteristics on a chosen day. By examining the isobars (the lines connecting points of equal atmospheric pressure) on a weather map, they could determine the direction, speed, and fetch of the wind. Maps from two or more days allowed them to determine the duration of the wind as well.

They then introduced the concept of "significant wave height," which they defined as the average (root mean square) height of the highest third of a set of waves. These waves appear to carry most of the energy in a stormy sea, and this term is now commonly used when discussing wave heights. As Munk explained in an interview (Finn Aaserud, La Jolla, Calif., June 30,1986):

> I think that we invented that [term], and it came about as follows: after returning to Scripps we were monitoring practice landings under various wave conditions by the Marines at Camp Pendleton, California. After each landing we would ask the coxswains of the landing crafts to estimate today's wave height, 7 feet one day, 4 feet the next day. We would make simultaneous wave records, and compute root mean square [rms] elevations. It turned out that the coxswains' wave heights far exceeded twice the rms elevations. It was easier to define a new statistical quantity than to modify the mindset of a Marine coxswain, so we introduced "significant wave height" as being compatible with the Marines' estimates. To our surprise that definition has stuck till today.

Next, Sverdrup and Munk focused on the rate at which the wind transmits energy to waves with significant heights. With nothing better in hand at the time, they adopted Harold Jeffreys's "sheltering" theory, which held that the wind would push on the windward side of a crest and leap over the next trough. The difference in air pressure from front to back of a crest would amplify a wave.

Then they formulated two energy balance equations, in which energy gained from the wind was balanced by the growth of wave heights and of wave speeds. One equation was valid when the fetch was effectively infinite and the wind blew for a limited time. The other equation was valid when the duration of the wind was very long but the fetch was limited. The two men assumed that, depending on the fetch and the duration of the wind, the wave heights would reach a steady state. From their theory they could then estimate the significant wave height and period. Next, they calculated how these waves escaped the storm area as swell and estimated how much the swell decayed because of air resistance as it traveled long distances toward a coast.

Their research was ready for use a few months before the invasion of North Africa and later Normandy in June 1945. As Munk recounted (CBS 8, San Diego, Calif., Feb. 18, 2009): "Eventually in collaboration with the British Met Service a prediction was made for the Normandy landing, where it played a crucial and dramatic role. The prediction for the first proposed day of landing was that it would be impossible to have a successful landing. As I understand, the wave prediction persuaded Eisenhower to delay for 24 hours. For the next day the prediction was 'very difficult but not impossible.' Eisenhower decided not to delay the second day, because the secrecy would be lost in waiting two weeks for the next tidal cycle." Because Sverdrup and Munk's information resulted in a delay of the invasion until June 6, when the swell was tolerable, their forecasts certainly saved many lives.

During the war Sverdrup and Munk also prepared charts and tables to guide amphibious landings elsewhere in Europe and Africa and taught young forecasters how to use them. Although many of their assumptions have since been superseded, their emphasis on energy balance laid the foundations of the science of wave forecasting and stimulated much postwar research.

The First Generation of Forecasting Models

In the 1950s Willard Pierson, Gerhard Neumann, and Richard James introduced a more rigorous approach to forecasting. They based their scheme on Munk's concept of energy balance but also introduced wave spectra and statistics. Pierson, a professor at New York University whom we met in chapter 5, developed a formal mathematical description of wave generation and propagation in 1952. To simplify the problem he introduced the concept of the "fully developed sea," in which the energy input of the wind would be exactly balanced by losses due to a number of causes. With certain strong assumptions, his team could calculate the shape of this final equilibrium wave energy spectrum. The spectra they derived for winds of different speeds have the peaked shapes we saw in figure 5.1. Wave frequencies around a "significant" frequency contain the most energy, while other higher and lower frequencies tail off rapidly in energy.

From such spectra the three scientists could derive the significant wave height and period of interest to mariners. In fact, assuming a bell-shaped distribution of wave heights, they could determine the fraction of waves that had any height of interest. They were also able to estimate these quantities when the wind is limited either in fetch or in duration.

These three authors published a manual for U.S. Navy forecasters written in simple, clear language for ordinary seamen. It presented the novice forecaster with a set of charts and tables with which to predict wave heights and periods in a high wind with a minimum of calculation. A forecaster only needed to know a few characteristics of the prevailing wind, such as its speed, fetch, and duration, to deal with any situation in the deep ocean. A test of their method showed agreement with wave height observations to only 50%—not high precision, it is true, but a step in the right direction. Their manual was used for over 20 years to design structures on vulnerable coasts and to avoid the worst of storms at sea.

The success of the manual was somewhat surprising in retrospect. Pierson and colleagues had made the basic assumption that the *shape* of the fully developed wave spectrum depends solely on wind speed and does not depend on wind fetch or duration. They assumed they could correct for limited fetch, for example, by cutting off a fully developed spectrum at a specific long wavelength. In order to test these assumptions and improve the accuracy of the method, new observations would be necessary and a more detailed theory would be required for the generation of storm waves.

More Progress in Forecasting

In 1955 Roberto Gelci, a marine forecaster at the French National Weather Service, independently conceived the energy balance approach to forecasting. Like Pierson, he proposed that the wave energy spectrum be calculated by balancing energy gains and energy losses. But he avoided the assumption of a steady-state, fully developed sea. He would instead try to calculate the evolution of the spectrum.

His scheme assumed that each wavelength would gain energy from the prevailing wind at a particular rate fixed by an empirical formula and would lose energy at some other rate. Gelci had no theory to guide him (this was 1955), so he had to extract empirical formulas for energy gains from crude archival observations. In addition, he could only assume that waves lost energy by spreading away from the wind direction. That would require him to calculate the paths of many different wave trains in order to predict their evolution. The final result was hardly a polished theory. At best, it was a recipe for calculating the changes in space and time of ocean waves in a strong wind.

With all the uncertainties entering his method, Gelci's predictions were far from accurate. But the idea of including the directions of wave propagation was novel, and many other investigators saw its potential.

Forecasting from First Principles

In the 1950s there was still a great deal of controversy about which physical processes were involved in raising waves on a flat sea. The situation improved in 1957 when Owen Phillips and John Miles published their resonance theories of wave generation. As you will recall from chapter 3, Phillips predicted a constant initial rate of growth from a flat sea; Miles predicted exponential growth beginning with small waves.

Both men tried to extend their theories from these "capillary" waves to predict the final equilibrium spectrum of gravity waves in a stronger wind. Phillips was able to predict at least the shape of the high-frequency tail of the spectrum by assuming that waves with frequencies near the peak of the spectrum reach energy equilibrium. But a comparison with observation showed that these theories predicted growth rates 10 times too small. So the theory of wave generation still remained more art than science, despite a flurry of theoretical attempts to improve the underlying physics.

Despite the success of the Pierson-Neumann-James manual, the reality of a fully developed sea was debated vigorously in the 1960s. Does the sea ever reach this kind of equilibrium? it was asked. Are both the fetch and the duration of a steady wind ever large or long enough to ensure a fully developed sea?

In 1964 Willard Pierson and Lionel Moskowitz claimed that they could find many examples of a fully developed sea in the records of ships at sea. Moreover, they showed that the shape of the spectrum of such a sea depends solely on the wind speed and that all the spectra can be transformed into a universal shape. They are "self-similar," as the Soviet scientist S. A. Kitaigorodskii had predicted. This demonstration helped to establish the Pierson-Moskowitz universal spectrum as a valuable working tool for forecasting. One needed to know only the wind speed, they argued, in order to predict the wave spectrum and therefore the significant wave height and period.

The Second Generation of Wave Forecasting

These developments marked what one could call the first generation of fore-casting models. Then in 1960, Klaus Hasselmann entered the forecasting field. At that time he was still a physics student in Hamburg, working on wave resistance to ships. After he read some papers by John Miles and by Owen Phillips on the transfer of energy among waves, he decided to investigate for himself. As we have already seen in chapter 4, he showed that energy trans-fers could occur only among sets of four waves that are related in frequency and direction. He claimed that such transfers could be a dominant process in wave growth, but without empirical verification, his claim remained contro-versial. Hasselmann also doubted that a fully developed sea (with a final steady energy spectrum) would ever be achieved in nature: the spectrum would con-tinue to evolve, even in a constant wind, because of the transfer of energy among waves.

The 1968 JONSWAP campaign in the North Sea, organized by Hasselmann (see chapter 5), was a turning point in oceanography and the science of fore-casting. It yielded high-quality observations of the wave spectra at a series of fetches and wind speeds. It also revealed that the changes in the shape of the energy spectrum with increasing fetch were "self-similar," which was a clue to the underlying physics.

The critical result for forecasting was that the empirical JONSWAP spectra could all be described by one universal empirical formula, in which the wind speed and the fetch are adjustable parameters. If one knew the wind speed and the fetch in a hurricane, one could predict the storm's wave spectrum by a simple modification of the universal formula. These test results were mark-edly different from the Pierson-Moskowitz spectrum of a "fully developed sea," where wind speed was the only necessary factor. The JONSWAP spec-trum contained four times the energy of the Pierson-Moskowitz spectrum at the same peak frequency. That would be crucial in estimating potential storm damage.

Moreover, the JONSWAP campaign revealed that nonlinear transfers of energy among waves were essential in the growth of waves, just as Hassel-mann had claimed. The wind deposits its energy to waves in the midrange of wavelengths, and these transfer energy to waves with longer and shorter wavelengths.

Energy Balance Models

The JONSWAP spectra are observational data that can be summarized by a universal formula, but they do not, in themselves, reveal the underlying physics that produces them. Oceanographers wanted to understand the forces that produce the observed spectra. Only then could they devise a theoretically sound forecasting scheme.

As with all other forecasting methods, their basic tool was the energy equation that balances the energy gains and losses of a specific wavelength as a function of time. However, they expanded the sources of energy gains and losses: a wave gains energy both from the wind and from other waves. It loses energy by spreading, by whitecapping, and perhaps by other mechanisms. To make progress, it would be necessary to find formulas that describe each of these processes. One way to start would be to try to extract trial formulas from the observations.

In effect, that is what happened. Using the JONSWAP data, oceanographers like Hasselmann tried out different approximations for the energy gains, losses, and transfers in the energy equation to see which combinations seemed to best fit the data. Then throughout the 1980s, they devised a variety of numerical forecasting models using the various approximations. This burst of activity initiated the second generation of forecasting models. One of the first examples is the Spectral Ocean Wave Model (SOWM) that Willard Pierson and colleagues constructed for the U.S. Navy. It was first applied to the Mediterranean Sea and later to the Atlantic and Pacific Oceans.

In 1981, a conference (the Sea Wave Modeling Program, or SWAMP—scientists love to create droll acronyms) was held to compare the performance of nine models in realistic exercises. Each model used the same input of wind conditions. The results were sobering: for a specified hurricane wind, the predictions of significant wave heights in the different models varied from 8 to 25 m—not a very comforting outcome.

Two of the weakest links in the models were identified: the approximations used for the dissipation of waves due to whitecapping and for the transfer of energy among waves. Hasselmann had written down the complicated formal mathematics for the transfer effect, but the computers of the day were unable to evaluate the effect within the time constraints of a daily forecast. Therefore, each expert adopted a different approximate formula for the effect, with the result that the models produced very different predictions. Hasselmann and colleagues would work furiously over the next decade to find a suitable approximate formula for the critical wave-wave interaction.

On to the Third Generation of Forecasting Models

The third generation of models began with the formation in 1984 of the Wave Modeling Group (WAM) under the leadership of Klaus Hasselmann. Over the next decade these researchers strived to produce a forecasting model that would avoid arbitrary choices of a limiting spectrum and would incorporate the best practical formulas for the wave-wave energy transfer. They also improved predictions of the spreading of storm waves away from the direction of the wind. The guiding principles of these third-generation models were that first principles must be used, rather than empirical "fitting" of data; that the wave energy spectrum would be created from these first principles, rather than have an assumed shape; and that the resulting nonlinear equations had to be solved explicitly, and not approximated.

The 1968 JONSWAP experiment had been fetch-limited and therefore could not establish whether a fully developed sea exists in nature; in the early 1980s the issue was still controversial. Therefore, in 1984 the Dutch scientist Gebrand Komen and other members of the WAM group performed a critical test of the latest WAM model (3GWAM). They wanted to see whether any tuning of the parameters of the model would result in a fully developed sea. They used a realistic model of the wind profile, a crude description of whitecapping, and better modeling of the spreading of waves. They found that indeed a steady-state spectrum is possible but is very sensitive to the way in which waves spread.

By the mid-1980s weather agencies and navies in several countries began to develop their own wave forecasting computer programs based on the energy balance concepts in the third-generation models. For example, the University of Technology in Delft, Holland, created the WAVEWATCH I program. It went through two revisions at the U.S. National Centers for Environmental Prediction and was eventually adopted as the standard forecasting tool. Similarly, the U.K. National Center for Ocean Forecasting developed the full Boltzmann forecast program, and the Canadians created the Ocean Wave model. Not to be outdone, the U.S. Navy's Fleet Numerical Meteorological and Oceanic Center published its own third-generation programs.

In the 1990s, satellite observations of wave heights and winds began to be available. Forecasters made heroic efforts to incorporate real-time satellite observations in their numerical prediction schemes. However, the interpretation of satellite radar images turned out to be a formidable task.

Daily Forecasts with Advanced Models

Third-generation forecasting programs require huge computer resources and became feasible only with the advent of supercomputers in the 1990s and early 2000s. At that point several nations could collaborate in producing daily or hourly forecasts tailored to specific areas in the global ocean. In the mid-1990s six European nations established a continent-wide center at Reading, U.K.—the European Centre for Medium-Range Weather Forecasts (ECMWF). Drawing on the pooled weather observations of member nations, the ECMWF issued forecasts for the continent and surrounding waters. Part of its charter was the twice-daily prediction of wave heights in the North Atlantic using third-generation models. By 2007, 34 nations had joined the ECMWF collaboration, and wave height predictions were being made for most of the world's oceans.

The U.S. National Weather Service has taken an independent path. It currently issues forecasts of weather and wave heights every 6 hours for seven large regions in the Western Hemisphere: the Western North Atlantic, the Caribbean, the Central Pacific, and the central and eastern parts of the North and South Equatorial Pacific. Predictions of wave heights are made with the updated WAVEWATCH III program.

As an example, here is an excerpt from the forecast for June 19, 2011, 1630 hours UTC.

HIGH SEAS FORECAST FOR MET AREA IV

1630 UTC SUN JUN 19 2011

SEAS GIVEN AS SIGNIFICANT WAVE HEIGHT . . . WHICH IS THE AVERAGE HEIGHT OF THE HIGHEST 1/3 OF THE WAVES. INDIVIDUAL WAVES MAY BE MORE THAN TWICE THE SIGNIFICANT WAVE HEIGHT

NORTH ATLANTIC NORTH OF 31N TO 67N AND WEST OF 35W

GALE WARNING

.INLAND LOW 48N65W 1002MB MOVING E 10KT. FROM 38N TO 50N BETWEEN 57W AND 60W WINDS 25 TO 35 KNOTS SEAS TO 9FT.

.24 HOUR FORECAST LOW 47N60W 1000MB. FROM 39N TO 51N BETWEEN 50W AND 58W
WINDS 25 TO 35 KNOTS SEAS TO 12FT.

.48 HOUR FORECAST LOW 47N53W 1004MB. CONDITIONS DESCRIBED WITH LOW 44N49W
BELOW.

Hindcasting: Comparing Forecasts with Actual Conditions

Forecasting models, like palm readings, are valuable only if they produce reliable predictions. There is no way to know which models, if any, make accurate predictions without comparing them with independent observations. To test their models, forecasters play a game called hindcasting. It consists of trying to predict the heights and periods of waves in a past storm using wind observations made during the storm. Then they compare their hindcasts with observations of waves made during the storm. Literally dozens of such exercises have been carried out with varying results. We'll examine a few of them.

The most pressing need for good forecasting arises during the hurricane season in the Gulf of Mexico. The National Hurricane Center in Miami, Florida, tracks these great storms with satellites and radar and attempts to forecast the path and strength of the winds and waves when the hurricane hits the coast. The U.S. Gulf Coast is densely populated with permanent tethered buoys that provide a continuous record of winds and waves. These observations have been compiled in a huge database that supplies forecasters with the input and ground truth for hindcasting.

In 1988 the WAM Development and Implementation (WAMDI) group carried out one of the first hindcasting tests with a third-generation model (3GWAM). They "predicted" the significant wave heights and periods for six North Atlantic storms and three Gulf of Mexico hurricanes: Camille (1969), Anita (1977), and Frederick (1979).

As an example of the kind of challenge these forecasters faced, consider Hurricane Camille. She was a category 5 storm when she entered the Gulf of Mexico on August 16, 1969. Her maximum sustained winds may never be known because she destroyed all the wind instruments when she made landfall. However, estimates at the coast were near 200mph. Even 75 miles inland, sustained winds of 120mph were reported. Six offshore oil drilling platforms recorded a maximum significant wave height of 14.5m (47.6ft), a value not expected to be exceeded within a century. When Camille landed at Pass Christian, Mississippi, the storm surge reached 24 feet. The hurricane dumped 10 inches of rain on the coast and a total of 30 inches in Virginia, with catastrophic flooding. With over a billion dollars of property damage and 259 deaths, Camille devastated the Gulf Coast with an impact that was not exceeded until Katrina.

For input to their hindcast the WAMDI group could use only partial wind observations before and during the storm. But they could compare their predictions with measurements of wave heights by a string of buoys off the Louisiana coast. The agreement was excellent right up to the peak of the storm, when the buoys were torn from their moorings and observations ceased.

The 3GWAM model also performed well for Hurricanes Anita and Frederick and even better for the six storms in the North Atlantic. The only question was whether accurate predictions could be made, say, 6 hours in advance. During a real storm, forecasters would be limited by the rapid changes in and accuracy of wind data, and by the power of their computers. Nevertheless, these excellent results motivated several nations to adopt third-generation forecasting models like 3GWAM and to provide nearly continuous forecasts.

Between 1988 and 2007 major improvements were made in forecasting models and in the availability of accurate wind observations by satellite radar. The 3GWAM model, for example, now had several variants, including one for relatively shallow coastal areas and another with an additional limit on the final spectrum. The Canadian program OWI was now in its third revision. These models were also tested by hindcasting some extreme weather conditions. R. Jensen and colleagues hindcasted six hurricanes, including some of the most notorious: Camille, Lili, Ivan, Dennis, Katrina, and Rita. Again, the matches with observations were very impressive.

Model against Model

How well do the different national weather centers predict wave heights in comparison with each other and for how long in advance? Do they agree and are they accurate? To find out, J.-R. Bidlot and M. W. Holt of the EC-MWF carried out a massive "validation" exercise in 2006. They compared the predictions of wave heights from February 1 to March 31, 2004, by six national centers who shared the same inputs of wind speed and direction.

The predictions were compared with observations made by a fleet of 79 buoys off both coasts of the United States and surrounding the British Isles. In one example they compared predictions made one day in advance by the U.S. National Centers for Environmental Prediction (using WAVEWATCH III) and by the U.K. Meteorological Office (using the Automated Tropical Cyclone Forecast). The ATCF (a second-generation model) performed just as well as the third-generation model but nonetheless was replaced by WAVE-WATCH III in 2008.

In figure 6.1 we see predictions of wave height by two different models one day in advance. The contours show the scatter of measurements. The 45-degree line going from the lower left to the upper right represents 100% correlation between measurements and predictions. As can be seen, the correlations are not quite perfect: for the scattered dots above the line, the predicted wave heights were slightly higher than the measured wave heights, and vice versa. Also the predictions are somewhat less accurate the higher the wave. But overall, the correlations are still quite close for both models. That was a general conclusion: these third-generation forecasts are reasonably accurate one day in advance.

Perhaps the best way to evaluate the progress forecasters have made is to ask how well they can predict wave heights as much as three days in advance. Bidlot and Holt found that the error in predictions increases from 0.5 m one day in advance to about 0.8 m at three days in advance. Not too bad!

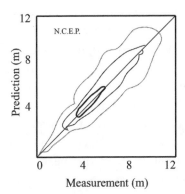 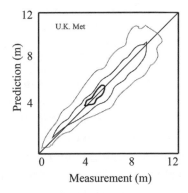

Fig. 6.1 Predictions and measurements of wave heights by two forecast centers, February 1–March 31, 2004. (Drawn by author from J.-R. Bidlot et al., "Intercomparison of Forecast Systems with Buoy Data," *Journal of the American Meteorology Society,* Apr. 2002, p. 287.)

Peter Janssen, a scientist at the ECMWF, carried out a similar study in 2007. He compared predictions of the continually upgraded 3GWAM model with buoy observations for the 14 hurricane seasons between 1992 and 2006. He found that the mean error in the predicted significant wave height increases, as one might expect, the more in advance the prediction is made. One day in advance, the mean error was always around 0.5 m; ten days in advance the error was still only 1.5 m. That's impressive.

Storm Surges and Surge Models

Forecasting maximum wave heights at sea during a hurricane is difficult, but estimating the storm surge at the moment a hurricane touches land is even more demanding. Storm surge is the hill of water pushed ashore by hurricane winds. The height of the surge at any point on land depends not only on the meteorological properties of the hurricane but also on the shape of the coastline it approaches. Gradually ascending coastlines (small slopes underwater close to shore) have the tendency to pile up more water onshore, leading to higher storm surges. Coastlines shaped like funnels tend to concentrate the waters, increasing their heights and their destructive potential.

At least five factors influence the formation of a storm surge. First, there is the low atmospheric pressure under the hurricane, which tends to suck up the surface of the sea, particularly under the eye of the storm. A 1-millibar drop in atmospheric pressure causes a 1-cm rise of the sea. With normal atmospheric pressure set at 1,013 millibars and hurricane eyes down to 909 millibars (Camille), that calculates out to over 100 cm, or 1 m, in surge height.

Second, the horizontal force of hurricane wind scoops up even more water and drives currents in the water. The currents tend to veer away from the direction of the wind because of the Coriolis effect. These unpredictable currents, turning to the right in the Northern Hemisphere and to the left in the Southern Hemisphere, greatly complicate the forecaster's task.

A third factor are the powerful wind-driven waves riding on top of the surge. Like all wind-driven waves, they don't actually move much water toward the shore, but when they break as surf, they carry considerable momentum. Their water can ride up the beach to a height above the mean water level equal to twice the height of the wave before it breaks.

As noted above, the shape of the coastline and the offshore bottom also influence the behavior of the surge. A steep shore, like that in southeastern Florida, produces a weaker surge but more powerful battering waves as they rise sharply up the beach. A shallow shore, like that in the Gulf of Mexico, produces a higher surge and relatively weaker waves.

Finally, if the hurricane reaches the shore near the lunar high tide for the area, the storm surge will be that much higher.

With all these factors acting simultaneously in a rapidly changing storm, forecasters have a daunting task to meet their responsibilities. Specialized models such as SWAN (Simulating Waves Nearshore) and SLOSH (Sea, Lake, and Overland Surge from Hurricanes) have been developed to meet the need for models of storms that impact coastal areas. SWAN was created by a team at the Technical University of Delft; SLOSH was developed at the U.S. National Ocean and Atmosphere Agency (NOAA) to model surges from hurricanes.

The SWAN Model

SWAN is a typical third-generation model that predicts the evolution of the two-dimensional wave spectrum by solving the energy equation, taking into account energy gains, losses, and transfers. In addition, it can be adapted to incorporate the shoaling, refraction, and diffraction that will occur at a specific location like the coast of Louisiana or Mississippi. It can even deal with the reflection and transmission of waves by breakwaters and cliffs.

The real-time performance of SWAN was tested in relatively mild conditions in 2003. In that year, the National Science Foundation and the Office of Naval Research sponsored the Nearshore Canyon Experiment (NCEX) near La Jolla, California. This shore is marked by two deep submarine canyons which greatly modify incoming swells from the Pacific. The goal of the experiment was to match real-time predictions of the onshore waves with buoy observations.

During the experiment, a 17-second swell approached the shore at an angle. The SWAN model predicted the variation of wave heights along the shore very well, as well as the offshore currents, but its predictions of wave heights were too low: a mere meter at best. In another NCEX exercise in the Santa Barbara Channel, real-time SWAN predictions of wave heights were too low again. Hindcasts showed that the presence of the Channel Islands upset the predictions.

The SLOSH Model

SLOSH was developed at the National Hurricane Center in Miami, Florida. It too is a third-generation computer model that takes into account the pressure, forward speed, size, track, and winds of a hurricane. Hindcasts show its predictions of surge heights to be accurate to within ±20%, not bad for such a complex model.

As an example of the use of SLOSH, the following forecast was issued by the Hurricane Center on August 27, 2005, a day and a half before Hurricane Katrina made landfall: "Coastal storm surge flooding of 15 to 20 feet (4.5–6.0 m) above normal tide levels . . . locally as high as 25 feet (7.5 m) along with large and dangerous battering waves . . . can be expected near and to the east of where the center makes landfall." Compare this forecast with the surges that Katrina actually produced: a maximum storm surge of more than 25 feet (8 m) at Waveland, Bay St. Louis, and Diamondhead. The surge at Pass Christian in Mississippi was one of the highest ever seen, with a height of 27.8 feet (8.5 m).

Because of the uncertainty in track forecasts, the SLOSH storm surge model is run for a wide variety of possible storm tracks. These possibilities are reduced to families of storm tracks, each representing one of the generalized directions of approach (west, north-by-northwest, north, northeast, etc.) that a hurricane would logically follow in a given area. These forecasts clearly improve in accuracy as the hurricane approaches the coast.

In a hurricane on the Gulf Coast, the highest surges occur on the right (east) side of the storm, where the winds are highest owing to their counterclockwise rotation. This side is often called the "dirty side" of the storm. This does not mean that the lower winds and surges on the left side of the storm are not dangerous: they may still cause massive destruction of the kind witnessed in New Orleans, which took the hit from the "clean side" of Hurricane Katrina.

Satellites in Forecasting

If you log onto the Internet at the NOAA WAVEWATCH III page of the National Weather Service's Environmental Modeling Center (as of this writing, http://polar.ncep.noaa.gov/waves/index2.shtml), you can see daily maps of wave heights and winds in all the world's oceans. You can also see forecasts as animated maps that change in steps of 3 hours, starting from when you sign on, to 180 hours in the future. This kind of information, freely available to the public, has been available only since about 2000. It has become possible because of two parallel developments: the improvement in forecasting models such as WAVEWATCH III and 3GWAM, and the advent of radar-equipped satellites. Each development stimulated improvements in the other.

Since about 1980, earth-orbiting satellites have revolutionized meteorology, oceanography, environmental science, and geology. Researchers are now able to observe the entire globe, in many wavelengths and in all weather, with minimum delay. Weather satellites enable us to observe hurricanes in all their fury and provide the raw data to allow us to predict the weather for the coming week. Ocean-monitoring satellites measure surface winds, currents, and wave heights in near real time. Some satellites measure the long-term variations of sea levels to within millimeters. Environmental satellites track deforestation in the Amazon basin, the temperature of the seas, and the melting of glaciers. We are far richer in our knowledge of our planet than we were only a couple of decades ago.

Ocean-Monitoring Satellites

Satellite observation of the oceans began with SEASAT in 1978. This satellite was equipped with several radar devices to image the sea. At that time many scientists doubted that radar could yield sharp images because of the random motions of the waves. But SEASAT proved them wrong. Postprocessing of its data yielded the first radar images of waves longer than about 30 m in a 10 × 15 km area. Unfortunately, SEASAT's electronics were crippled after 3 months by a short circuit, and the satellite was abandoned.

At present four satellites are monitoring the oceans, sending down a flood of data. Several nations have banded together to build and operate these satellites and share their data. The European Space Agency launched ERS-1 in 1991, ERS-2 in 1995, and ENVISAT in 2002. The United States and France collaborate to operate JASON-1 (launched in 2001) and JASON-2 (2002).

ERS-1 and ERS-2 are equipped with several types of radar altimeters as well as instruments to measure ocean temperature and atmospheric ozone. They have collected a wealth of data on the earth's land surfaces, oceans, and polar caps and have been used to monitor natural disasters such as severe flooding or earthquakes. ENVISAT is an improved and updated version of ERS-2.

JASON-1 and -2 are more specialized, being dedicated to observations of the oceans. They carry a suite of radar altimeters to measure wave heights and surface winds. In addition, JASON-2 measures variations of sea level as small as a millimeter over a year, information that can be used to follow the impact of climate change on the oceans. This precision allows the hills and valleys in the ocean surface to be mapped; from these data, the ocean currents can be predicted. JASON-2 also monitors the El Niño effect and the large-scale eddies in the ocean. Both versions of JASON continue the pioneering observations of Topex/Poseidon, which operated between 1992 and 2006.

Each of these four ocean-monitoring satellites orbits the earth in about 100 minutes, passing from pole to pole. Moreover, their orbits are *sun-synchronous*, meaning that the plane of each orbit always faces the sun by continuously turning around the earth at the same rate as the earth revolves in its orbit around the sun. That ensures that a given point on the earth is observed at the same local time every day. That is useful, for example, in measuring sea temperatures.

The satellites map the sea in strips that are spaced apart in longitude, so that it takes several days for a satellite to pass over the same point on the earth again (for example, ENVISAT takes about 3 days to do so). A global network of tracking stations continually monitors the satellites, so that their positions in space are known within centimeters. The stations also download data from the satellites periodically.

Data analysis and archiving takes place at multiple forecast centers. In the United States and the United Kingdom, the active agencies are NOAA and the Meteorological Office, respectively, both of which use the WAVEWATCH III forecasting model. The European Union's ECMWF uses the 3GWAM model.

Satellite Radar Magic

Each ocean-monitoring satellite is equipped with three radar instruments: a *scanning radar altimeter* to measure wind speed and direction and wave heights; a *scatterometer* to measure wind speed near the ocean surface; and a *synthetic aperture radar* to measure directional wave spectra. These radars use microwave radiation to measure the distance of the satellite from the tops of the waves, accurate to within a few centimeters.

As you probably know, radar (the word originated as an acronym for "radio detection and ranging") was invented during the Second World War in a collaboration between British and American scientists; it is generally credited with saving the British from invasion by Nazi Germany. The *radar altimeter*, which is the principal instrument on board the ocean satellites, was actually invented way back in 1924 by Lloyd Espinschied, an engineer at Bell Telephone Laboratories. He took a number of years to make it practical for aircraft, however. It wasn't until 1938 that it became standard equipment on commercial planes. SEASAT carried the first set of satellite radar altimeters in

1978. To appreciate how radar altimetry produces useful observations of the sea we need to get into the weeds a bit.

The basic idea behind radar is simple. A radar antenna emits a burst of short pulses of microwave radiation. (By "short," I mean measured in nano-seconds, or billionths of a second; by "microwave," I mean radiation with wavelengths of a few centimeters.) The same antenna that emits the pulse receives the echo of that pulse from the target after a time delay that depends on the distance to the target. The distance is equal to the delay multiplied by the constant speed of light, 300,000 km/s.

Let's say the antenna is pointing straight down from the satellite. Then the radar performs as a simple altimeter, recording the present distance of the satellite from the sea. But unlike an aircraft altimeter, which converts the measurements into heights from the ground *up*, a satellite altimeter con-verts the measurements into distances from the satellite *down*. This is pos-sible and desirable, because the present height of the satellite (say, above mean sea level) is known independently to *within a few centimeters* from the tracking stations that monitor the satellite in its orbit. That means that the radar distances to the sea can resolve the heights of waves above mean sea level.

In practice it is the *shape* of the return pulse that indicates the mean heights of the waves. If the sea were perfectly flat, it would act as a mirror and the leading edge of the return pulse would rise very sharply to its peak inten-sity. If, on the other hand, the sea were rough, the slope of the leading edge would be gradual, as waves of different heights reflect the incident pulse.

With some modifications an altimeter becomes a scatterometer, which measures the wind speed at the sea surface. It is based on two empirical re-lationships: (a) wave heights increase with increasing wind speed; and (b) the taller the wave, the more it scatters microwave radiation. Here is how it works: When the sea is rough, the incident pulse is scattered in many direc-tions by the random faces of the waves. Therefore, the return pulse is some-what weaker than if the sea were flat and were acting as a mirror to reflect all the incident radiation. This *decrease* in the intensity of the return pulse, beyond that expected for a flat surface, can be calibrated to yield the speed of the wind.

A simple altimeter on a satellite yields wave heights along a *line* in the sea. A side-scanning altimeter sweeps a pencil beam back and forth in the direction perpendicular to the satellite's path, like a blind man sweeping his cane back and forth as he works his way down a sidewalk. These sweeps build up a two-dimensional image whose spatial resolution (100–300 m) is determined by the width of the radar beam.

Synthetic aperture radar (or SAR) also sweeps out a two-dimensional strip of the ocean but with spatial resolution much better in its forward direction than is possible with a scanning altimeter. For example, the European ocean satellite ERS-1 delivered a radar map of the waves in a 5 × 10 km area at intervals of 200 km along the satellite's path. It achieved a resolution of 30 m in the forward direction and 100 km in the transverse direction. The reason it can achieve this higher resolution is that it illuminates each patch in the target area repeatedly at many different angles of incidence, with pulses of a fixed microwave frequency. Each return pulse from the patch is Doppler-shifted in frequency because of the changing angles of incidence. The Doppler shifts change predictably as the satellite passes over the patch, so these shifts effectively label the patch's location in the target area. The satellite's electronics use the label to add together the intensities of all the echoes from that patch, producing a crisp, high-resolution image.

Figure 6.2 illustrates the principle of SAR. Imagine a satellite moving in a precise orbit that serves as a reference frame from which distances to the sea can be measured. The straight line labeled V (for velocity) is the satellite's path in the figure. At regular intervals along the path (marked by the dots), the satellite emits a radar pulse at a fixed radio frequency—say, at 13 gigahertz (GHz) (2.3-cm wavelength). For our purposes we can say that the radar beam has the shape of a pyramid, whose base is a rectangular footprint on the sea. Each pulse travels to the sea, where the beam strikes the surface, illuminates the waves there, and is reflected.

Each ocean wave in the rectangular footprint of the beam reflects a part of the return pulse, and its part is Doppler-shifted in frequency, relative to the incident frequency of 13 GHz. The Doppler shift depends on the location of the wave in the footprint. The shift is proportional to the component of the satellite velocity along the line of sight from the satellite to the wave. So for example, the shift is zero for a wave directly under the satellite; it is negative (blue-shifted) for waves further toward the front edge of the rectangular footprint and positive (red-shifted) for waves toward the rear edge. In this way the position of every wave in the footprint is labeled with a unique Doppler shift.

As the satellite moves from position 1 to position 2, each wave in the footprint is illuminated repeatedly, and each time its distinctive Doppler shift

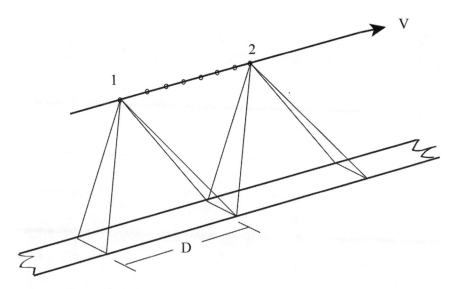

Fig. 6.2 Illustration of the principle of synthetic aperture radar on a satellite moving from point 1 to point 2. *V*, satellite's path; *D*, length of beam's footprint on the sea. (Drawn by author from S. W. McCandless, Jr., and C. R. Jackson, chap. 1 of NOAA, *Synthetic Aperture Radar Marine User Manual*, 2004.)

changes in a predictable fashion. The satellite electronics use the Doppler label to collect and add the intensities of all the echoes from a given wave in the footprint. In addition, the electronics process all the distance measurements to this wave to determine the height of its crest. In practice, many thousands of echoes are collected for each wave in the footprint.

The result is a high-resolution three-dimensional microwave image of the sea along the path of the satellite. Figure 6.3 shows an example, an image of San Francisco Bay and its famous bridge. It was made by the Jet Propulsion Laboratory's experimental aircraft, with novel radar that uses polarized beams for higher spatial resolution. (I have enhanced the waves for greater clarity.) Individual waves longer than about 30 m show up with a brightness that depends on the wave's height.

This image now has to be processed to obtain the spectrum of wave heights from which the significant wave height can be determined. It is this wave height that appears in the forecasts for the region. It can be compared with the significant wave height that is calculated with, say, the WAVE-WATCH program, given the local wind as input.

Fig. 6.3 A microwave image of San Francisco Bay in the vicinity of the famous bridge. (Courtesy of Yijun He et al., *Journal of Atmospheric and Oceanic Technology* 23 [2006]: 1768, used with permission of the American Meteorological Society.)

SAR works because the radar pulse travels at the speed of light, which is much faster than the satellite's speed or, for that matter, faster than any changes in wave shapes. I have skipped over the nasty complications involved in processing such SAR data using sophisticated Fourier-based algorithms. For example, I have assumed that the Doppler shifts are precisely correlated with a wave's position within the footprint. But the waves are moving randomly, and these motions introduce random Doppler shifts (noise) in the coding of the waves as the thousands of radar transmissions are sent and received. The result is a slightly fuzzier image than would be produced if the waves were absolutely stationary.

Radar has been honed into a superb tool for imaging the ocean. Much thought has gone into extracting useful information from SAR data. Once again, Klaus Hasselmann and his close associates (including Sophia, his wife) were the leaders in this effort. The technique applies not only to surface waves but also to interior waves and the topography of the bottom, as well as the temperature of the sea and the direction and strength of surface winds. These modeling tools and satellite measuring techniques have been critical in understanding the different types of waves that we'll cover in the next chapters.

CHAPTER SEVEN

Breaking Waves

Extreme Waves for Extreme Surfers

Peahi Beach, on the north shore of Maui, Hawaii, is famous for some of the largest breaking waves in the world. Several times each winter, storms near Alaska produce swells that cruise thousands of kilometers to crash as surf at Maui. About half a kilometer offshore lies a deepwater reef, and when a 10-m swell hits the reef, the waves rise to heights of 20 m and more.

It's hard to describe how powerful and majestic these waves appear to an observer on the beach. Each one seems to rise slowly, implacably out of the off-shore swell. Then there is that spine-tingling moment when the crest hovers on the brink of breaking. As the crest finally begins to curl over at the highest point of the wave, it forms an enormous tunnel of air. This tunnel propagates rapidly along the front of the wave. And then the wave collapses in a thundering mass of foam and surges far up the beach. What a thrill you get just by watching!

But the ultimate thrills are experienced only by the madcap crew that come to surf these monsters. They are members of the exclusive club of "extreme" tow-in surfers, who arrive every winter to compete in riding the highest possible wave and setting a world record. These surfers need to be towed by Jet Skis beyond the breaking waves because the waves move too fast (48 km/h) to catch by paddling. Indeed, the waves are so awesome and so dangerous that the beach was nicknamed "Jaws" by the surfers who first discovered these immense breakers.

How high a wave is it possible to surf? Professional surfers travel the world to find out. Over time, as their skills and daring—and surfing technology—have improved, they have tackled taller and taller waves. Back in 1969 Greg Noll was credited with surfing the highest wave ever ridden at Makaha on the western shore of Oahu, Hawaii. It was "only" 10 m high.

But by 2001, tow-in surfing had made it possible to reach the bigger waves, and the record had jumped quickly to just under 20 m. Mike Parsons, a Californian surfer, traveled to the Cortez Bank, 160 km off the San Diego coast, to find a higher wave. The bank is an underwater mountain range that rises to within a meter of the surface. When a high swell hits the bank, huge waves are created. On January 19, 2001, Parsons caught a fabulous wave and set a new record: 20.1 m (66 ft). He received an award of $66,000 for the feat, the largest prize ever won by a professional surfer. Parsons was 36 years old at the time, a ripe old age for a surfer.

Parsons's record didn't last very long. On January 10, 2004, Pete Cabrinha, another professional, set a new world record of 21.3 m (70 ft) at Jaws, Maui. He earned $70,000 for a few minutes of work and for risking his life. Mike Parsons was not about to be overtaken, however. On January 5, 2008, Parsons broke his own record at Cortez Bank by riding a wave estimated at more than 70 feet. Then later in 2008, Parsons set the official world record at Cortez Bank by riding a wave 23.5 m (77 ft) high. That's as tall as a seven-story building. "I couldn't believe it was that big. The drop just never ended. It went down, down, down, down," Parsons said in an ESPN interview after his ride.

Nonetheless, back in 1998 Ken Bradshaw was credited unofficially with surfing an 80-foot wave (24.4 m) at Outer Log Cabins on the north shore of Oahu, Hawaii. Could others beat that record?

Parsons's official world record of 23.5 m remained secure for a decade. Then on November 9, 2011, Garrett McNamara was videotaped surfing a wave off the coast of Nazaré, Portugal, where an underwater canyon focuses high swells to impressive heights. His monster wave was rated unofficially at a jaw-dropping 27.4 m, or 90 feet (fig. 7.1 shows him exiting the wave). Then on January 28, 2013, McNamara may have broken his own record with an estimated 100-foot wave, also at Nazaré.

In December 2011, a board of expert judges was convened to examine the videos of McNamara's November 2011 ride. Using McNamara's shin bone as a length scale, the judges concluded the wave was actually "only" 25.7 m (78 ft) high. It was not 90 feet, but it was still a world record, topping the old mark by a whole foot. McNamara's wave was accepted by Guinness World Records in May 2012. In an interview on ESPN Action Sports, the 44-year-old McNamara said he had been on a mission for the previous 10 years to catch the "biggest, best waves on the planet." McNamara's feat will be hard to beat. Who will ever ride a certified 80-foot wave?

Fig. 7.1 Garrett McNamara exiting his record-breaking 78-foot wave. (Photo 27758072, dreamstime.com.)

Swells Become Surfing Waves

Giant surfing waves are created when a big swell, sometimes as high as 10 m, meets some offshore obstruction like a steep slope or a reef. So to understand breaking waves we need to begin with the life cycle of a swell.

You'll remember that swells are the remnants of the chaotic waves in a distant storm. As the wind blows over a long distance (the fetch) for several days, it creates tall waves with a broad spectrum of wavelengths, sharply peaked crests, and a wide range of directions. For example, a storm with winds of 50 knots (93 km/h) that blow for several days over a fetch of 1,000 km could generate waves 10 m high. After the winds die down, the longest waves— with a period of, say, 15 seconds and a wavelength of 350 m—could escape the disturbed area at a speed of 84 km/h.

A swell like this can travel thousands of kilometers with hardly any decay. As it rides into shallower water near an island or continent, the swell is transformed, changing shape, speed, and height. Let's see how this happens.

Refraction

As a swell approaches a shore, it is affected by the varying depth of water over the bottom. One effect is a change in the direction of the swell. On our very first walk along the beach in chapter 1, we noticed how the wave fronts of a swell turned to face nearly parallel to the shore. This effect is an example of *refraction*, the bending of a wave front due to a variation of wave speed along the front.

In water shallower than half a wavelength, the speed of a wave decreases with decreasing depth of the bottom. So if, for example, the left side of an incoming wave front passes over shallower water than the right side, the left side will slow down and the wave front will turn toward the left.

At an actual beach or bay, where the sandy bottom may have a complex shape, incoming waves may be turned in different directions. For example, in the top drawing of figure 7.2, we see a headland that extends as an underwater ridge some distance from the shore (the dashed lines are contours of the bottom). Incoming waves will turn to climb the ridge. They are focused by the changing depth of the water on the flanks of the ridge. The opposite effect occurs in a bay with a concave bottom, as shown in the bottom drawing. In both of these cases, the waves tend to shape the areas on which they break: over time, the bay becomes more semicircular while the point of land gradually erodes away from the relentless pounding of the focused waves.

Shoaling

A change of direction is not the only change a swell incurs in shallow water. On the approach to shore, the swell grows taller (a process called shoaling) before it eventually breaks. How does this happen?

A simple explanation for the growth of a wave as it approaches the beach is that the wave preserves its rate of energy transport until the instant it breaks. Its energy transport rate, according to Airy's theory, depends on the square of its height and on its speed. As we saw just now, a wave slows down in shallow water. So if the energy transport rate remains constant and the speed declines, the height must increase.

Another way of looking at shoaling is to imagine that the swell is riding up a steep, smooth slope toward the beach. The wave's period would remain constant, but the wave speed would decrease as the wave encountered shallower water. Each crest would move slightly slower than the crests behind it. As a result the crests would bunch together, like a line of cars approaching a stalled vehicle on a highway. Therefore, the wavelength (the distance between crests) would decrease as the wave approaches the shore.

Now let's look more closely inside the wave to understand why the wave grows taller. When the wave is in water deeper than half its wavelength, the blobs of water under the surface rotate in synchronized circular orbits,

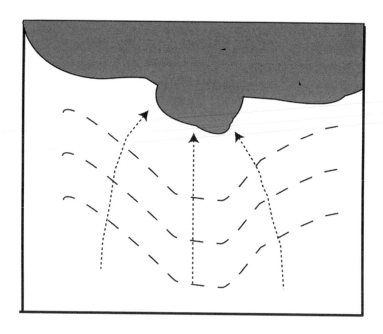

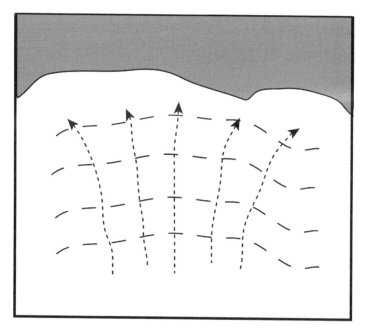

Fig. 7.2 Shoaling waves are refracted by the varying depths of water near the shore. In this illustration the long dashes show the contours of depth; the short dashes show the paths of waves. Incoming waves turn toward a headland that extends underwater (*top*) and diverge in a concave bay (*bottom*).

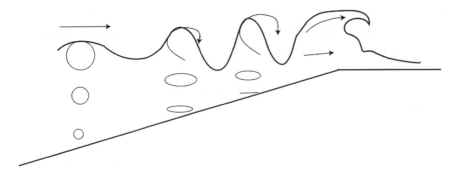

Fig. 7.3 As a swell moves into shallow water, its underwater orbits change size and shape. The wave's height increases, as does the orbital speed of the topmost orbit. When the orbital speed exceeds the wave's phase speed, the wave will topple over.

returning to nearly their original positions on each circuit (see fig. 2.3). The greater the depth, the smaller is the diameter of the orbit.

However, as the wave rides up the slope, the orbits change shape from circular to horizontally elliptical (fig. 7.3). At a depth of about half a wavelength, the deepest orbits touch the bottom, and their vertical motions are inhibited. The orbital blobs no longer trace out perfect circles in one place; rather, they trace elliptical orbits that start moving toward the shore. In still shallower water, the deepest orbits flatten into virtually linear orbits, in which blobs oscillate only horizontally. The energy they lose from their vertical oscillations is transferred to orbits closer to the surface (if we ignore friction). These more energetic top blobs now trace out increasingly larger ellipses as they move further and faster toward the shore. In effect, the kinetic energy within the wave shifts upward and forward.

As you can see in figure 7.3, the orbit closest to the surface determines the height of the wave—that is, the distance from crest to trough. As this orbit gains energy from its neighbors, it cannot spin faster because its rotation period is a fixed fundamental property of the wave. Therefore, the orbit absorbs additional energy by increasing both its vertical and horizontal dimensions. Hence, we get a towering wave just before it breaks.

The Breaks

Now, when does a wave break? One answer is that the wave becomes unstable when its steepness ratio (height to wavelength) is about 1 to 7. So when a wave with a wavelength of 7 m reaches a height of 1 m, it becomes unstable. Another criterion is the slope of front face of the crest: when it becomes vertical, the wave crashes. A rough rule of thumb, valid for very gentle slopes of the bottom, says the wave will break when its height reaches 80% of the depth of the water. But why doesn't such an unstable wave break backwards out to sea? The dynamic reason for breaking forward is the *acceleration of the crest*. Here's how it happens.

As described earlier, the topmost orbit of a shoaling wave absorbs energy from its lower neighbors. Consequently, its diameter increases while its rotation period remains the same. That means the water on top of this increasingly large orbit has to increase its forward linear speed. But the top water forms the crest. So the crest moves forward faster and faster than the water below it. At some point the crest overtakes the trough ahead of it and tips over. In other words, the orbital speed of the water in the crest becomes faster than the wave's phase speed (the wavelength divided by the period), and the crest topples over.

In real life, friction with the bottom is not negligible. When a wave's deepest orbits touch the bottom, friction will slow the base of the wave but not the crests. Therefore, the wave is sheared and will eventually break. The bottommost orbits still do move forward, albeit slower than the top, so that their contact with the bottom moves the sediment gradually toward the beach. Coastal engineers are vitally concerned about the transport of sand, the evolution of beaches, and the development of currents, especially during winter storms. We'll return to this matter in a moment.

Breakers: Spilling, Plunging, and Surging

After riding in toward the shore and building in height, a wave does finally break. The shape of the breaking wave depends primarily on the slope of the bottom but also on the wave's steepness. There are three main types of breaking waves: spilling, plunging, and surging.

Steep waves moving onto gentle slopes (those with, say, a rise of 1 m in 100 m) become *spilling* breakers (fig. 1.2, top), in which water tumbles from the crest down the front face as a beard of white foam. The wave's height decreases slowly over a long distance as the wave moves toward the shore and dissipates its energy in turbulent froth. These breakers can give a surfer a nice, long, but undramatic ride.

Moderately steep waves riding on long, moderate slopes (say, a rise of 1 m in 20 m) generate *plunging* breakers. Alternatively, an abrupt change of slope, such as an underwater reef that faces deep water, can produce plunging breakers. Their crests curl over and fall forward of the front face (fig. 1.2, bottom). A spectacular splash or jet may result from the impact, and a tube of air may be trapped under the falling crest. These are the tubes (or barrels) in which daring surfers love to ride. Big barrels are called Mackers: you could drive a Mack truck through them.

The perfect wave for surfers is a wave that develops its tube by curling over progressively along its crest. The wave shown at the bottom of figure 1.2 is curling over from left to right. This happens when the wave is approaching the shore at a slight angle to the slope of the beach. If the wave were to hit the slope dead on, the whole crest would curl over at once, making it dangerous or impossible for surfers to ride.

A wave with low steepness that encounters a steep slope (say, a rise of 1 m in 10–15 m) can become a *surging* breaker. Its crest doesn't crumble or pitch over as it approaches shore. Instead, the wave remains intact as it rides up the slope, until suddenly the front of the wave collapses all at once and the water surges far up the beach. Surging waves have steep, smooth, fast-moving front faces.

Surging waves can be dangerous because they don't look threatening. A pair of newlyweds learned that on March 7, 2011. They were having their wedding portraits videotaped on the beach of Bodega Bay, California, and were posing with their backs to the sea. The bride was wearing her long white dress and veil; the groom was dressed more casually. Without warning, a big wave broke as a surging breaker and knocked them off their feet. Happily, they escaped with nothing more than a good soaking. As compensation, their video went "viral" on TV, and they enjoyed a touch of celebrity for a while.

Predicting the Worst Breakers

Coastal engineers earn their living by designing the wharves and seawalls we see around a harbor. In order to build a safe structure, an engineer has to be able to predict the biggest breaker that can be expected to arrive at the structure. Until recently, the theory of breakers was inadequate to make reliable predictions. Therefore, engineers had to rely on observations near a coast or on laboratory experiments in a wave tank. From such data they were able to extract empirical rules, but without a clear understanding of the physics underlying the rules.

For instance, they devised an empirical breaking index to predict what type of breaker will form on a particular slope. The index, termed *Xi*, is dimensionless and equals the product of the steepness (height divided by wavelength) and the slope of the seabed. Here are some examples of the index:

Xi value	*Type of breaker*
>3.3	Surging
0.5–3.3	Plunging
<0.5	Spilling

We'll see that numerical simulations have refined these criteria.

Then in 1974, J. Richard Weggel (Coastal Research Center, Washington, D.C.) compiled a useful design tool for the coastal engineer. After analyzing all the available observations on breaking waves and extracting empirical formulas from the data, he used these formulas to create a set of graphs to help engineers estimate the maximum wave heights of breakers on a structure near the shore. His practical results are still in use today.

Specifically, Weggel found that the maximum wave height of a breaker at a structure depends on the depth of the water at the structure, the steepness and period of the incident wave, and the slope of the bottom. As an example, consider a breakwater in 20 feet of water which is fronted by a bottom with a slope of 1:20. How high will breakers form at the breakwater from a swell with a 10-second period? Dipping into Weggel's graphs we find a maximum breaker height of 0.83 times the 20-foot depth of the water, or 16.6 feet.

Weggel wrote that "breakers higher than 16.6 feet will break further offshore from the structure and will have dissipated a sizable fraction of their energy before reaching the structure. While smaller breakers may reach the structure they will not exceed a critical design condition" (*Maximum Breaker Height for Design*, U.S. Army Coastal Engineering Research Center, 1973). Thus, the slope seaward of a structure acts as a filter for the wave spectrum, causing higher waves to break further offshore.

Modeling a Breaking Wave

Thirty years after Weggel published his work, engineers are still relying on empirical rules to predict maximum wave heights at particular locations. Several experimental groups have continued to explore how waves break, however. For example, W. Alsop and his team at the Wallingford Laboratory in Oxford, U.K., use a long computer-controlled wave tank and sophisticated instrumentation to measure waves breaking on smooth slopes. They have obtained more reliable and detailed relationships among the wave parameters.

As powerful computers, sophisticated computer algorithms, and improved modeling techniques have become more available, researchers have tended to replace physical wave tank experiments with numerical simulations. These computer models are cheaper, faster, and more flexible (as well as drier) and can be used to test current knowledge of the physics of breaking waves. A good example is the 1997 investigation of Stephan T. Grilli, I. A. Svendsen, I. A. Subramanya, and R. Subramanya at the University of Rhode Island.

Grilli was born in Belgium and received his training in civil engineering and oceanography at the University of Liège. He joined the University of Rhode Island in 1987 and eventually became chair of the Department of Ocean Engineering. He has specialized in coastal engineering and has become expert in numerical modeling of dynamic fluid behavior. Since 1998 he has been studying tsunamis caused by underwater landslides. Following the 1994 Indian Ocean tsunami he organized an expedition to obtain images of the seafloor at the epicenter. The team used a remotely operated vehicle to photograph the rupture zone at a depth of 4,500 m.

Grilli's team calculated the evolution of a solitary wave as it shoals and breaks over a mild or moderate slope. The wave arrives in deep water with an assumed initial depth at the foot of the slope. The computer model, which contains the two-dimensional, nonlinear equations of hydrodynamics, calculates the changes in the shape and speed of the wave, taking into account the gradual decrease of water depth. Figure 7.4 shows a typical example of a plunging wave. The curling lip at the front is realistic even though the model does not assume any friction on the slope.

From numerous trials with these models, using different slopes and initial wave heights, Grilli was able to define a breaking criterion that predicts the type of breaker—spilling, plunging, or surging—that will appear. He found that the criterion is proportional to the slope and inversely proportional to the square root of the ratio of initial wave height and initial water depth. Translated, this means that the various types of breakers are associated with the breaking criterion as follows:

Breaking criterion	Type of breaker
0.3–0.37	Surging
0.025–0.3	Plunging
<0.025	Spilling

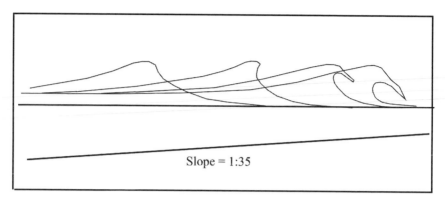

Fig. 7.4 A numerical simulation of a breaking wave. The wave is shown at successive stages as it climbs a slope of 1:35. (Drawn after S. T. Grilli et al., *Journal of Waterway, Port, Coastal, and Ocean Engineering,* May/June 1997, p. 107.)

This result provides engineers with even more accurate ways of predicting the types and strengths of waves that might hit breakwaters and other structures.

Grilli was also able to determine how the maximum height of a breaking wave depends on the slope. Contrary to intuition, the flatter the slope, the higher the maximum height and the deeper the water where the wave breaks. So, for example, on a shallow slope of 1:100 (i.e., a 1-m rise over a 100-m distance), a wave reaches twice its initial height and breaks at a depth equal to 40% of the initial water depth. On a steeper slope of 1:15, however, the wave rises to only 1.4 times its initial height and breaks at a depth of one-seventh the depth at the foot of the slope.

Grilli's model has an impressive ability to follow the development of the curling lip of a plunging breaker, including the air tube underneath the curl. In examples published in 2003, Grilli and coworkers tracked the impact of the curl on the water and the splash that follows. These results encourage us to think that the basic physics is reasonably well understood.

There is still much more to be explored with numerical simulations, and several groups have embarked on ambitious projects. For example, Qun J. Richard Zhao and his team at the University of Delaware are working to include turbulence in their models of breaking waves. P. Higuera and a Spanish team at the University of Cantabria are investigating the movement of gravel on a slope as a wave shoals and breaks. Their results so far compare favorably with laboratory results. Oceanographers at Delft University in the Netherlands have also upgraded the sophisticated SWAN (Simulating Waves Nearshore) computer program that was discussed in chapter 6. SWAN now takes into account bottom friction and whitecapping, wave diffraction around obstacles, and refraction due to currents and depth. An extended three-dimensional version of the program (Delft3D) can also forecast currents and sediment transport.

Beach Currents Caused by Waves

When a wave breaks, the water under it surges forward as surf, a foaming mass of moving water. All the energy contained in the wave is dissipated in turbulence and in the kinetic energy of the water. The last, dying gasp of the wave is a thick layer of water that rides up the beach in the "swash zone" to a high point, stops, and drains back to the sea. This backwash water has to go somewhere, so it flows under or between the incoming waves in a system of currents. Oceanographers recognize three main types of beach currents: rip, undertow, and longshore.

Rip currents flow in narrow channels out from the shore. These channels are typically trenches through sand bars or along jetties. In order to drain the water that constantly arrives at the beach, these narrow rip currents must be very fast, with speeds that range from 0.5 m/s to over 2.5 m/s (9 km/h). They can easily pose a threat to a careless swimmer, who can be carried swiftly out to sea with no chance of swimming against the rip. But because the rip currents are narrow, the smart swimmer knows to turn parallel to the shore and swim sideways out of the channel, where they can again turn to swim with the surf to the beach. Rip currents are actually stronger on their surfaces, and so do not tend to pull swimmers underwater.

As the name indicates, an *undertow* is a broad deep current that returns the excess of water directly under the breaking waves back to the sea, rather than in narrow channels. Undertows are slower than rip currents as a rule because they flow in broad paths back to the sea. They can still be dangerous to weak swimmers, but they are not nearly as threatening as rip currents.

Most waves approach the shore at an angle. When they break, they push water along the shore as a *longshore current* that carries a heavy load of sediments—primarily sand, but also pebbles, gravel, and other debris. At some point along the shore, the turbulent current dies out, and the load starts to drop. The flow patterns depend upon the wind direction, the wave strength within the swash zone, and the type and strength of the currents flowing back to the sea.

Longshore currents can transport sand long distances along the shore. If waves arrive from different directions relative to the shore, they can generate longshore currents in opposite directions. Thus, large quantities of sand may oscillate along the shore, sculpting the beach.

Beach Cycles

A beach may grow, move, or disappear entirely because of wave action. If, for example, waves erode a cliff at one end of the shore, the rubble may be carried by a longshore current for some distance and deposited in a bare place. A new beach will build up at this place. Or if the source of sand is another beach, that beach may be transported along the shore to a new location.

A beach normally goes through a seasonal cycle. In summer, gentle waves move sand from offshore and drop it on the beach in a thick layer. A flat terrace, called a "berm" may be formed up the beach. Then in winter, high storm waves erase the summer berm and deposit its sand in a winter berm higher up the beach. At the same time the strong backwash of the winter waves pulls sand off the lower part of the beach and builds a sand bar offshore. The beach as a whole is left bare or covered with gravel. This cycle can maintain a beach in oscillating equilibrium for many years.

Some beaches are strongly affected by storm-driven waves, especially waves with a very long fetch. These "storm beaches" usually have very steep slopes (up to 45 degrees) composed of rounded cobbles, shingle, and only occasionally sand. As might be expected, these components are sorted along the slope of the beach. The smallest pebbles lie within the swash zone, while the largest cobbles lie in berms at the top of the beach. The ferocious storm waves wash away any sand, while the backwash trickles through the larger cobbles rather than washing them away. A number of such beaches can be found on Cape Hatteras, North Carolina, and on the western coasts of England and Scotland.

Beach Cusps

Breaking waves can leave a permanent pattern in the sand of a beach. One distinct pattern is called beach cusps, but a better name might be chain of arcs. Each arc has two "horns" that point toward the sea and a bow that points up the slope of the beach. The horns of neighboring arcs touch in "cusps." An arc measures a few meters to perhaps 60 m between horns. Each arc in the pattern has about the same width and height.

The arcs are formed as water from a breaking wave sloshes up the beach slope and drains back to the sea. These cusps are self-sustaining—that is, once they have been formed, they maintain themselves. As waves crash first on the horns, they start slowing down, first dropping the heavier pebbles at the horn and then splitting in half to roll into the two arcs on either side of the horn. As the wave enters into the arc, it slows down even more and eventually collides with the half wave that was split at the next horn down, causing the remaining finer sediments to fall out until remnants wash up exhausted on the swash zone. Finally, the backwash drains downslope at the center of the arc in a kind of rip current.

Two theories have been proposed to explain the formation of these interesting cusp patterns: self-organization and standing offshore waves. In self-organization, a slight initial depression in the sand is enhanced because it attracts and accelerates the flow of water, which erodes the top of the arc. Once formed, the pattern maintains itself. Alternatively, a system of standing waves in the near offshore region may produce a variation in the height of the slosh along the shore, which creates the chain of arcs. Despite numerous simulations and field studies, investigators have not been able to demonstrate a clear choice between these theories. The subject is not purely academic because it bears on beach erosion and repair. And beach cusps appear on beaches around the world, so studying them is useful.

Tracking Waves across the Continental Shelf

In recent years oceanographers have carried out a number of campaigns to measure waves and currents on a coast and to improve forecasting models such as SWAN and Delft3D. A favorite site for such work is on the North Carolina coast. Off the Outer Banks town of Duck, the continental shelf rises slowly and smoothly toward the shore for about 100 km. It provides an ideal outdoor laboratory. The Army Corps of Engineers has maintained a coastal research station there for many years.

In the summer and fall of 1997, the Corps hosted several hundred scientists in a field experiment called Sandy Duck. Their goal was to better understand how sediment is moved by breaking waves and how beaches evolve. The scientists deployed a large array of buoys and sensors along the shore as well as along a line perpendicular to shore. In addition, they set up an array of fixed sensors in the surf zone to measure currents and sediment concentration. They carried out 30 different experiments, ranging from the measurement of waves, currents, and bottom topography, to the swash of broken waves up the beach.

During this campaign a major storm that raged for four days increased the maximum wave height to 4 m. During similar storms in 1985 and 1990, a sandbar had formed and moved offshore, leaving a deep trough behind it. As each storm passed, a system of rip channels had torn through the bar. But in 1997, a long bar formed without a deep trough close to the beach. The source of the sand remained a mystery. Moreover, the bar contained rip channels throughout the whole period rather than just after a storm. Several explanations were proposed for this behavior, but it became clear that too much was still unknown.

The coast at Duck is battered by hurricanes every year from September to December. A team from the Naval Postgraduate School at Monterey Bay, California, decided to monitor the waves at Duck during the hurricane season of 1999, as part of a larger exercise called the Shoaling Wave Experiment (SHOWEX). They set up a line of six waverider buoys between depths of 21 and 195 m. These buoys measure horizontal and vertical displacements at the sea surface; from these measurements, wave height spectra and directional spectra can be determined. In addition, the team laid down six bottom pressure sensors in a line parallel to the shore at a depth of 24 m to measure wave directions accurately.

The Navy team was extremely lucky. In three months they experienced four hurricanes—Floyd, Gertrude, Irene, and Jose (categories 5, 4, 3, and 2, respectively). Their equipment worked continuously for three months without a flaw. They scooped up a ton of data.

During Hurricane Floyd, the significant wave height reached 8 m at 100 km offshore but only 4 m near the shore. A similar effect was seen during Hurricane Gert. Thus, the wave energy decayed by a factor of 4 as the waves crossed the inner shelf. Sonar images of the sandy bottom showed the cause: a pattern of rough, sandy ripples. Evidently, the rough bottom interfered dramatically with the propagation of the waves. This result raised an intriguing possibility: that the movement of sand affects the movement of waves as they approach the shore, just as the waves affect the sand. In other words, there is a dynamic interaction involved in the erosion and buildup of a beach.

Sandbars

For many years scientists at Duck had assumed that the behavior of waves, sand, and currents was the same anywhere along the straight, smooth shoreline. But recent research demonstrated that in fact, waves at two locations separated by only a mile had a very different behavior. What could account for this? Several explanations were proposed. Perhaps the geology of the shore varies more than has been realized. Or perhaps the size of sand grains varies along the shore. Or perhaps the drainage properties of the shore differ from one location to another.

Britt Raubenheimer and her team from the Woods Hole Oceanographic Institution decided to test a different but related idea: are the locations of underwater sandbars the prime factors in determining where erosion occurs during a storm? When a sandbar forms offshore, it could shield the beach from storm waves. In contrast, a sandbar near the shore could allow waves to flood the beach and cause erosion of berms and cliff faces. The scientists went to Duck in 2000 and measured waves and mean water levels as storms attacked the shore. The data strongly supported the sandbar explanation, but several other factors still needed to be explored. Only new experiments at a more complex site could decide the issue.

The Nearshore Canyon Experiment

By the fall of 2003, the community of nearshore scientists was ready to mount a massive attack on the problems of nearshore dynamics. Two famous research organizations, Woods Hole Oceanographic Institution and Scripps Institution of Oceanography, agreed to coordinate the efforts of three dozen scientists, engineers, and students from a dozen institutions. They were now equipped with more advanced technology as well as improved forecasting tools such as SWAN and Delft3D. For their Nearshore Canyon Experiment (NCEX), they chose two submarine canyons off the California coast at La Jolla. The site, close to Scripps, had been studied by Scripps scientists in earlier years. This time an all-out effort was planned.

The heads of these canyons are only a few hundred meters off the shoreline. Their complex bottom topography strongly disturbs the persistent swell from the Pacific and generates strong surf, reflected waves, and currents. Conditions vary dramatically along the coast. At Black Beach the waves are reliably high enough to give surfers a good ride. Just two miles down the coast at La Jolla, the waves are consistently gentle. So the primary goals of the NCEX campaign were to observe the shoaling waves and the currents they produce at different locations, to map the movements of sand, and to test the reliability of the Delft3D program.

As an example of the types of hardware deployed by the group, consider the efforts of a team from the U.S. Navy Ocean Waves Laboratory. They set up five waverider buoys to measure wave heights and speeds at the head of Scripps Canyon, where the strongest effects of the bottom topography were expected. A string of 17 pressure-velocity sensors was set up along the 10-m-depth contour north of the canyon, to measure alongshore currents outside the surf zone. A region just south of the canyon is completely sheltered from the direct swell by the submarine topography. To monitor this quiet region, an array of bottom pressure sensors was established. The team gathered data for a full three months, from September to December 2003.

Another team from the Naval Research Laboratory wanted to learn how sensitive the models of currents and surf were to the actual topography of the ocean bottom. They deployed a huge variety of instruments to gather new measurements. Aside from the usual waveriders and pressure gauges, they used autonomous underwater vehicles, airborne video cameras, commercial satellite imagery, digital motion imagery, and sonar equipment mounted on Jet Skis.

Britt Raubenheimer and her husband Steve Elgar, both scientists at Woods Hole, were primarily interested in how wave reflection and refraction redirect the flow of water in a complex canyon system. They observed how wave "setup"—the piling up of a giant hill of water during a storm—generates large alongshore and cross-shore currents. Variations of setup along the coast could account for much of the variation in current strength. But reflection and refraction of waves also affects the strength of currents. They also measured the momentum of onshore waves, the surge of water up the beach, the return of water in rip and undertow currents, and the transport of sediment. Their primary tool was an acoustic Doppler velocimeter—an instrument analogous to sonar. The device emits a stream of high-frequency sonic pulses focused at a single point, receives the echoes, and records the data. Twenty-five of these devices were deployed along 3 km of shore, at depths ranging from 2.5 to 15 m. The team gathered a mountain of data that will occupy them for years to come.

Several teams reported their preliminary findings at the fall 2004 meeting of the American Geophysical Union. A team from the U.S. Naval Research Laboratory, led by K. Todd Holland, had used an array of color digital cameras along 2 km of shore to monitor waves and surf out to a depth of 25 m. The images obtained were analyzed in real time with an automatic process to obtain hourly estimates of wave period, wave direction, breaker height, and the effects of bottom topography. Estimates were also made of directional wave spectra and surf flow speeds. These estimates were used to fine-tune the Delft3D forecasting program, which was also provided with the canyon topography and incoming swell heights. Delft3D's predictions were compared with the actual observations. The researchers commented that there were significant correlations between the model's predictions and the observations via video. Translating from science-speak, this was a very good result. The agreement was particularly good for rip currents. The team concluded that their method of incorporating constraining data at the boundaries of the site resulted in marked improvement in the accuracy and resolution of the predictions.

We can look forward to a time when oceanographers can predict with some accuracy the impact of storm waves on a complicated shoreline. This knowledge will allow coastal engineers to armor the coast to protect homes and lives.

CHAPTER EIGHT

Freaks and Rogues

For God's Sake, Hold On—It's Got Us!

"It's got us, boys!" Ernest Shackleton screamed. He and five crewmen were in a 20-foot lifeboat in the midst of an Antarctic gale, on a desperate 800-mile voyage. Now they were in mortal danger, tossed in mountainous seas and freezing in the icy Antarctic storm.

In 1915 Shackleton had led a 23-man crew on an expedition to be the first to reach the South Pole for the glory of England. But their ship, *Endurance*, had been locked in the sea ice and slowly crushed to matchsticks. Shackleton and his crew had no choice but to risk a perilous attempt to escape in two small boats. With luck, courage, and skill they managed to reach rocky Elephant Island, home to hundreds of elephant and fur seals. But they were still 800 miles from South Georgia, where lay a large whaling station. Shackleton decided to sail on to South Georgia with a small crew to seek help. Now in this horrendous storm, the end seemed near.

Shackleton later wrote:

At midnight I was at the tiller and suddenly noticed a line of clear sky between the south and southwest. I called to the other men that the sky was clearing and then a moment later I realized that what I had seen was not a rift in the clouds but the white crest of an enormous wave. During twenty-six years' experience of the ocean in all its moods I had not encountered a wave so gigantic. It was a mighty upheaval of the ocean, a thing quite apart from the big white-capped seas that had been our tireless companions for many days. I shouted "For God's sake hold on! It's got us!" Then came a moment of suspense that seemed to be drawn out into hours. White surged the foam of the breaking sea around us. We felt our boat being lifted and flung forward like a cork in breaking surf. We were in a seething chaos of tortured water; but somehow the boat lived through it, half filled with water, sagging to the dead weight and shuddering under the blow. We bailed with the energy of men fighting for life, flinging the water over the sides with every receptacle that came to our hands, and after ten minutes of uncertainty we felt the boat renew her life beneath us. She floated again and ceased to lurch drunkenly as though dazed by the attack of the sea. Earnestly we hoped that never again would we encounter such a wave. (Sir Ernest Shackleton, *South: The "Endurance" Expedition*, 1919)

Shackleton did make it to South Georgia and was able to rescue all of his men. Their epic voyage made maritime history—and that gigantic wave is part of that history.

Sailors' Tales

Stories like this have abounded in the folklore of the sea for hundreds of years. Sailors have often told of moments of absolute terror when rogue waves taller than their ship's mast rose without warning, barreling toward them like some monster in a nightmare. Sometimes these waves appeared suddenly out of a calm ocean; more often, they arose in a violent gale. And sometimes they came in groups of three, named the notorious Three Sisters: just when the petrified sailors thought they had survived the monster, two more giant waves would crash down on their lumbering boat in rapid succession.

Many of these accounts were probably exaggerated by the sailors' horror and maybe embellished for effect, but the sheer weight of eyewitness testimony lends some credibility to them. Paul C. Liu, a scientist at NOAA's Great Lakes Laboratory, listed several dozen of the most notable in the journal *Geofizika* in 2007 (vol. 24). Most of these occurrences are based on eyewitness reports and rough estimates of wave heights, but they do not lack for drama. For example, in 1896 Joshua Slocum was sailing around the world single-handedly in his 40-foot sloop *Spray*. He wrote this account of his experience much later, in his book *Sailing Alone around the World*: "One day, well off the coast of Patagonia, while the sloop was reaching under short sail, a tremendous wave, the culmination it seemed of many waves, rolled down upon her in a storm, roaring as it came. I had only a moment to get all sail down and myself up on the peak halyards, out of danger when I saw the mighty crest towering masthead-high above me. The mountain of water submerged my vessel. She shook in every timber and reeled under the weight of the sea, but rose quickly out of it and rode grandly over the rollers that followed. It may have been a minute that from my hold in the rigging I could see no part of the *Spray*'s hull."

Not Just the Small Boats Suffer

Small boats are by no means the only craft that have been threatened by giant waves. The luxury ocean liner *Queen Mary* encountered one in 1942. In December of that year, while carrying 16,000 troops from New York to Great Britain to fight in the war, this 82,000-ton ship was pounded by a storm in the North Atlantic. Some 700 miles west of Scotland she was hit broadside by a powerful wave that rolled her by an angle estimated later at 52 degrees. Another 3 degrees and she might have capsized. But she slowly righted herself and pressed on to her destination. Walter Ford Carter, whose father was aboard at the time, estimated that the wave was 92 feet (28 m) high. Another account placed it at 70 feet (21 m), but there was no damage to the ship that could support either number.

The wreck of the oil tanker *World Glory* provides dramatic evidence of another giant wave. On June 13, 1968, the 45,000-ton ship was fighting her way through heavy seas along the southeast coast of Africa, the scene of many a shipwreck. She was riding southward on the infamous Agulhas Current and facing into a fierce gale that was blowing against the current. The waves were 50 feet (15 m) high, very steep, and more than half the length of the 737-foot ship. Captain Androutsopoulos reduced speed to the minimum required to point the bow directly into the oncoming rollers. Nevertheless, the ship was swept with green water for hours.

Then suddenly, an extraordinary wave, estimated at 70 feet (21 m), rolled under the ship, lifted her high up, and let her hang for a moment. The captain heard an ominous crack and saw the bow tilt down at a frightening angle: the ship had snapped in half under its own weight. Within a few moments the two halves began to drift apart, still battered by high waves. Only a few survived to tell their tale at the inquiry.

There is no doubt that many, many ships have been crushed in giant seas in the midst of a brutal storm. What is at issue is the size of the largest waves, how frequently they appear, and what the state of the sea was. Waves 30, 40, even 50 feet high are sometimes seen in the North Atlantic during winter or in the "Furious Fifties" of the Southern Sea. But until recently, many oceanographers believed that sailors were exaggerating and that waves higher than, say, 50 feet were mere fantasy. There *are* some early incidents that give one pause, however. Consider, for example, the following.

In February 1933 the U.S. Navy's oil tanker *Ramapo* was trapped in a massive storm that stretched from Asia to the U.S. Pacific coast. The ship had delivered 70,000 barrels of fuel oil to the Naval Station in Manila and now was returning home to San Diego. Sixty-knot winds had been blowing from the west for seven days, across thousands of miles of ocean. Now the waves were huge—50 feet (15 m) from crest to trough. They rolled forward at 45 knots, driving the ship before them. In this following sea, the ship was pitching wildly, with its bow pointing skyward every 15 seconds.

At three o'clock on the night of February 7, the officer on the bridge happened to look back to the stern and saw a monster wave overtaking the ship. At that instant, the ship's stern was in a trough and the bow was tipped up at a sharp angle. In one brief look, the officer saw he could line up the crest of the wave behind him with the crow's nest on the mast amidships. As he watched, the enormous wave lifted the stern, broke over the ship in a thundering mass of water, and drove it forward into a deep trough.

Incredibly, the ship survived, with only minor damage to the structures on deck. And thanks to the brief, almost unconscious observation of the officer, it was possible to estimate the height of the monster wave from the length of the ship (148 m) and the position of the crow's nest. The shocking result was a crest-to-trough height of 34 m, or 112 feet! It was the largest wave ever reported.

The crew of the *Ramapo* was lucky. If the ship had been longer than the 340-m wavelength of the wave, it might have been lifted amidships and cracked in half just like the *World Glory*. But as it was only about half a wavelength long, the ship rode the waves reasonably well. This eyewitness report has the support of a measurement, however imperfect it might have been in the moment of crisis.

Still more credible evidence emerged from the voyage of another luxury liner. In September 1995, the 70,000-ton *Queen Elizabeth II* was crossing from Cherbourg to New York. To avoid Hurricane Luis, the ship sailed south of its normal track. Nevertheless, the waves were topping 18 m. On September 11 at four in the morning, the windows of the Grand Lounge, at a height of 22 m (73 ft), were smashed in.

Eleven minutes later an enormous wave loomed out of the darkness. (See the preface for a photo of a similar wave.) Captain R. W. Warwick said: "It looked as though the ship was heading straight for the white cliffs of Dover. The wave seemed to take ages to arrive but it was probably less than a minute before it broke with tremendous force over the bow. An incredible shudder went through the ship, followed a few minutes later by two smaller shudders. There seemed to be two waves in succession as the ship fell into the 'hole' behind the first one. The second wave of 28–29 m whilst breaking, crashed over the foredeck, carrying away the forward whistle mast."

Warwick recalled later that he saw that the crest of the second wave was directly in the horizontal line of sight from the bridge, at a height of 29 m (96 ft). In addition, a Canadian buoy in the area registered a wave of 30 m (98 ft) sometime during that storm.

Broken windows are convincing evidence for the heights of waves. Take the example of the *Caledonian Star*, a cruise ship that was returning from a visit to Antarctica in March 2001. A fierce gale had been blowing for several days off Cape Horn, the tip of South America. On March 2 the ship was struck by a wave that smashed the inch-thick windows of the bridge, 98 feet above the sea. The officer at the helm said later that the wave was solitary and twice as tall as the average.

These enormous rogue waves are still taking lives despite new technologies and ship safety systems: in April 2012, the headlines in San Francisco read, "Search for four missing sailors called off after fatal rogue wave hit yacht during race" (Associated Press, April 15, 2012).

A Celebrity Rogue Wave

The most famous wave in maritime history was recorded by instruments on the *Draupner* oil rig in the North Sea off Norway on January 1, 1995. This wave is notable not only for its size but also for the fact that for the first time an objective instrument made the measurement.

A severe storm had been raging for several days. A down-pointing laser wave gauge, mounted at one of the corners of the platform of the oil rig, was recording continuously during 20 minutes out of every hour during the storm. The time series shows that the significant wave heights—the mean wave height, trough to crest, of the highest third of the waves—were generally around 11–12 m during the entire afternoon of January 1. From standard statistical arguments one could expect the maximum wave height (crest to trough) would be about 20 m.

Then at 3:30 p.m. a single wave spiked in the records with a maximum height (crest to trough) close to 26 m—about 85 feet! Not the biggest ever reported, but the source was unassailable, and the fact of being single makes this wave unique. Moreover, one would not expect a wave this tall to appear in a sea of 11-m waves very often. At the time, Paul Taylor, a scientist at Oxford University, estimated the chance of a recurrence was 1 out of 200,000 waves, or one wave in 10,000 years.

MAXWAVE: Studying Rogue Waves

The *Draupner* wave caused quite a bit of excitement among ocean engineers and oceanographers. It established without question that extraordinarily high waves do exist and supported visual estimates of their sizes. A search for more examples revealed that sensors on the North Sea oil rig *Goma* recorded 466 waves over 70 feet high in 12 years, or nearly one a week, thus refuting the oceanographers' disbelief of sailors' extreme wave stories. But were these waves abnormal rogues, or might they have been unusually tall waves resulting from violent storms?

In any case, giant waves were far more numerous than previously thought. And one search of marine casualty records showed that at least 20 supertankers and container ships (more than 200 m long) had been lost between 1969 and 1994 (Douglas Falkner, quoted in Graham Lawton, "Monsters of the Deep," *New Scientist* 170, June 30, 2001). Could rogue waves have caused any of these expensive wrecks? A thorough scientific study of the phenomenon was urgently needed.

First, it was critical to distinguish between a rogue wave and "normal" tall waves. In fact, a rogue wave is not necessarily very tall, although great height makes great headlines. A rogue is just *unexpectedly* tall, considering the state of the sea. A good engineering definition is a wave whose height is twice the significant wave height. So in a sea with 3-m waves, a single 6-m wave would still qualify as a rogue by this definition. It probably would appear suddenly and disappear without doing much damage to a ship. The key point is that nobody knew why and how these rogue waves developed, whereas "normal" tall waves were obviously created by violent storms.

Therefore, in December 2000 the European Union funded a three-year investigation of shipping accidents and their possible connection to rogue waves. The project, called MAXWAVE, was headed by Wolfgang Rosenthal

and Susanne Lehner, senior scientists at the GKSS Research Center at Gees-
thacht, Germany. Eleven meteorological and oceanographic institutions in
Europe joined the effort.

Rosenthal and Lehner outlined four major goals for the project: to confirm
the existence of rogue waves and estimate the risk of encounters, to apply new
knowledge of rogues to ship design, to develop forecasting methods for rogues,
and to disseminate the information to the marine community. The final goal
of this project, they wrote, was "to improve the understanding of the physical
processes responsible for the generation of extreme waves and to identify geophysical conditions in which such waves are most likely to occur" (*Journal of
Offshore Mechanics and Arctic Engineering* 130 [2008]: 21006). The project would
exploit, for the first time, the archives of satellite SAR observations to develop
reliable statistics and make detailed studies of individual events. As Rosenthal
and Lehner explained, the ESA satellites ERS-1 and ERS-2 produce radar images
of waves in a 5×10 km area every 200 km, with 30-m resolution. These images
are normally used to generate wave spectra that indicate the *average* conditions
in the scanned "box." But the raw SAR images contain more detailed information which could be analyzed for the possible appearance of rogue waves.

Rosenthal and Lehner's team (one of seven) obtained 34,000 images,
evenly distributed around the globe, for a 27-day period from April to May
2001. The images were processed and automatically searched for individual
extreme waves with a specially designed technique. Despite the relatively
small size of this data sample, the team discovered 10 extreme waves of 25-m
height. One wave had a height of 28 m (92 ft) from crest to trough.

In another study a 27-day series of ERS-1 images obtained during August
and September 1996 was processed to reveal maximum wave heights in 3×3
degree areas (about 180 miles square) over the whole globe. The resulting
map showed a belt of tall waves (18–20 m high) in the Furious Fifties latitudes
of the Southern Ocean and in the track of Hurricane Fran in the western
North Atlantic. No waves higher than 8 m were seen anywhere else.

These were useful results, but how extreme were these waves? More precisely, how often does a single exceptionally tall wave—say, one 20 or 25 m
high—appear, and what is the significant wave height at the time? Are these
waves to be expected as rare eruptions in a very violent sea? Or are special
physics needed to explain them?

To find out, the team examined the nearly continuous records of wave
heights obtained at the *Draupner* and *Ekofisk* oil rigs in the North Sea and off

the Belgian coast. The observations were made by down-pointing laser in-struments and floating buoys and had been collected over two decades. Their great advantage, in addition to their continuity, is that they could reveal the profile of each wave. From the data the team could determine the significant wave heights in a stormy or quiet sea, as well as the maximum heights of in-dividual waves.

The oil rig records showed that individual rogue waves like the one re-corded from the *Draupner* have a steep front; a sharp, pointy peak; and long, flat trough. The whole *Draupner* event, from the preceding trough to the fol-lowing trough, lasted a mere 10 seconds. So we can see how difficult it is to find and measure individual rogues.

The team was surprised to find that the frequency of waves twice the sig-nificant wave heights was *not* unusual. The distribution of wave heights was exactly what is expected for a randomly varying sea surface. A few exceptions were found, such as the famous *Draupner* wave, but on the whole the frequen-cies of wave heights, even very large heights, are normal. Very tall waves, the researchers concluded, are just rather rare. We shall see that this conclusion has been contested up to the present time.

Where and How Are Rogues Formed?

Do rogue waves occur preferentially at particular locations? At a news confer-ence on July 21, 2004, Rosenthal and Lehner claimed they do. Their study of three weeks of ERS SAR data suggested that rogues arise where waves meet currents, such as the Agulhas off the east coast of South Africa or the Gulf Stream. But they admitted that rogues can also appear far from currents, where weather fronts are moving. "We know some of the reasons for the rogue waves, but we do not know them all," Rosenthal concluded.

Another MAXWAVE team tried to find wind or wave conditions that would be likely to produce a rogue wave. They compiled a database of over 650 ship accidents in heavy weather between 1995 and 1999, using the Lloyd's Marine Insurance files. The data were pared down to 245 accidents for which locations and times were known. Then hindcasts were calculated to obtain wave spectra, significant wave heights, wave steepness, and several other parameters.

The study failed to find a smoking gun—that is, a reliable predictor for a rogue wave. In particular, the significant wave height on the day before or during these accidents was surprisingly low, about 1–5 m. So even if the tallest wave was 10 m, it still didn't follow that it caused the accident. Many accidents were, however, associated with *steep* waves, and some were associated with a *cross-sea*, in which a swell was crossing the wind direction at a large angle.

A map of the locations of the 245 accidents showed that they were concentrated in four areas of heavy shipping: the U.S. East Coast, the North Sea, the Mediterranean Sea, and the coast of China. This result led to the suggestion that ocean currents such as the Gulf Stream and Kirushiro Current could be generating rogue waves. The Agulhas Current, on the southeast coast of Africa, had previously been suspected because of its long history of shipwrecks. But no wrecks in the Agulhas were found in this study. Actually, the Cape of Good Hope, where the Agulhas meets the prevailing west wind, seemed a better candidate for producing rogues.

Although the findings of the MAXWAVE project proved inconclusive on the causes of rogue waves, researchers were able to summarize the observed properties of these extreme waves. They are at least twice the significant height in size, so they are statistically unexpected. They are very steep and peaked and follow an unusually deep trough. These waves can arise singly or in small groups, in calm seas or violent storms, in shallow or deep water, in coastal currents or far out at sea. They are fast and sweep by in at most a few tens of seconds.

At the end of the MAXWAVE project, a team of theorists proposed three classes of rogue waves, based on what they looked like and their profiles in space and time, along with possible mechanisms for their generation.

1. A single extreme wave—a *wave tower*—might arise from the interaction between a wave train and a current. When a wind-driven train encounters a current flowing in the opposite direction, the waves can pile up to an extreme height. The Agulhas Current was still considered a candidate for this type.

2. Rogue waves can occur in groups of three or more high peaks—the *three sisters* effect. Sister waves are thought to receive energy from the atmosphere by a line of thunderstorms preceding a cold front and traveling in the propagation direction of the group.

3. Two wave trains with similar wavelengths but crossing each other in different directions (a cross-sea) can produce an interference pattern that produces unusually tall waves. This condition is called a *white wall* and is probably what the *Queen Elizabeth II* encountered.

The MAXWAVE project was most successful in exploring the frequency of rogues and in devising techniques to locate them in SAR images. But much work remained to apply these techniques to the mountains of available SAR images. Therefore, Lehner and Rosenthal organized a follow-on program called WaveAtlas, which aims at producing global maps of individual rogue waves. As of mid-2011, a million SAR images from 1998 to 2000 had been processed. Each covers a small area, 5×10 degrees (about 300×600 miles). Several new hotspots for rogue waves have already been identified: the southwest coast of Greenland, the North Pacific in winter, the southwest coast of Australia, and Cape Horn, at the southern tip of South America. The analysis continues.

Tall Storm Waves

With all the current excitement about rogue waves, we need to be reminded that some very tall waves are not necessarily rogues. Tall waves can be expected in a particularly violent storm. Take, for example, the experience of a British research team in February 2000. This group boarded the 295-foot ship *Discovery* on February 6 and sailed west from Southampton. They planned to measure the wind speeds and wave heights at the peak of a powerful storm. The team had done this sort of thing before and expected no great difficulties. They were in for some surprises.

By February 8 the ship was 250 km west of Scotland, near the small island of Rockall. The wind was howling at a steady 72 km/h (45 mph). The captain of the ship headed straight into the wind, but nevertheless the ship rolled badly. During the next 12 hours the wind rose to 83 km/h, and the sea went mad. The team was measuring waves with a significant height of 18.5 m (60 ft), and these waves were setting a record. But the best—or worst—was yet to come.

Past midnight on February 8 the ship was battered by waves of 20, 25, and an astounding 29.5 m (96.8 ft), the largest ever measured with scientific instruments. Windows broke. The ship almost collided with a trawler. A lifeboat came loose and was banging against the starboard side. People suffered bruises and broken ribs.

As tall as these waves were, they were not rogues: they could be expected with such record-breaking significant wave heights. However, these monster waves did pose a problem: they continued to arrive regularly for 12 hours, long after the peak winds had subsided. So evidently these huge waves were not wind-driven; they were waves in a swell. The team is still struggling to find an explanation for them.

What Causes Rogue Waves?

After the MAXWAVE project published its results in 2002, rogue waves were accepted as real, with a much higher occurrence probability than 1 in 10,000 years, as Paul Taylor had posited. It was also clear that they were responsible for some shipwrecks. The next logical scientific step was to understand them in physical terms. Could scientists learn enough about their origins to be able to predict them?

Theorists have picked up the challenge enthusiastically. Indeed, modeling rogue waves has become a hot topic in the past decade. The first question that has to be settled is whether rogue waves are just very rare examples of a random wave or are the result of some special kind of physics.

What do scientists mean by a "random" wave? They think of the tossing sea surface as an interference pattern among many weak Stokes waves that move in different directions (see chapter 4 for Stokes waves). When several waves overlap constructively, they create a taller wave. But because they move independently, their meeting at one place and time is a random event, seemingly unpredictable. Most of the sea surface is covered with such random waves.

Random doesn't mean chaotic, however. There is method to the madness of waves because they obey the laws of chance. The most familiar example of a random process is the tossing of coins. If you toss a coin 10 times, you expect to get 5 heads and 5 tails, more or less, because the chance (probability) of getting either a head or a tail in each toss is the same—50%. But if you repeated this 10-toss game a thousand times, there is a good chance that at least once you would get 10 heads and no tails. That's an "extreme" event that is due entirely to chance. If you plot the difference between the numbers of heads and tails in a long series of tosses, you get a bell-shaped curve, the so-called Gaussian, or "normal," distribution. The peak of the curve corresponds to equal numbers of heads and tails which you would get most of the time, and the wings of the curve correspond to rare ("extreme") events where there were far more heads than tails, or vice versa.

In the same way, 10 ocean waves might overlap and interfere constructively to build a towering wave 10 times higher than any of the original waves. This would be a rogue event, due entirely to chance.

Way back in 1880, John William Strutt (later known as Lord Rayleigh), of Cambridge University, gave a mathematical description of random waves. This was a relatively minor accomplishment in his vast output. He made im-

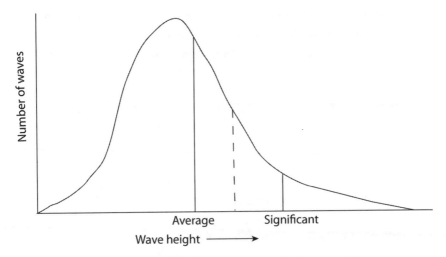

Fig. 8.1 Rayleigh distribution of wave heights in a random sea. The highest third of waves are found in the tail, up to the dashed line.

portant contributions to all the fields of physics, including optics, electromagnetism, and sound. He was also the first to provide a scientific answer to the age-old child's question "Why is the sky blue?" (The answer was molecular scattering of the sun's light waves.) He was above all a theorist, but in 1904 he shared a Nobel Prize with William Ramsay for the discovery of a new element, argon.

But getting back to his contribution to the understanding of ocean waves, he viewed the sea surface as made up of many overlapping sinusoidal waves. He also assumed that their frequencies and amplitudes were distributed according to a bell-shaped curve and had propagation directions and phases distributed uniformly. The waves could interfere with one another and produce a fluctuating sea surface. His was a theory of linear waves which are intrinsically weak but may combine to create tall waves.

Rayleigh derived a simple formula for the distribution of wave heights graphed in figure 8.1. The higher the wave, the more infrequent it is. The highest waves appear in the long, thin tail of the distribution. Rayleigh's predictions of wave heights agreed reasonably well with the observations of his day. But that was before rogue waves were accepted as real.

Rayleigh's theory of wave heights has been extended to include a broader range of frequencies; to deal with wave groups instead of infinitely long trains; and to treat nonsinusoidal waves with cusps and deep troughs. The question today is whether a Rayleigh-type distribution can account for the estimated frequency of rogues. Several tests suggest that it can.

For example, in 1998 marine scientist Michel Olagnon examined buoy measurements of 2 million waves and found no exceptions to the Rayleigh distribution. In 2000, he repeated the comparison with 3 million waves and confirmed his earlier conclusion. Then in 2008, five years of buoy measurements off the Mediterranean coast of Spain were analyzed by Leo Holthuijsen and colleagues. They had 10 million waves to work with, and found waves with heights as large as 2.8 times the significant wave height, but still the Raleigh distribution fitted the data nicely.

The North Sea is a proven breeding ground for extreme waves because of the frequency of powerful storms and the possible influence of a shallow bottom, less than 100 m deep in places. Oil rigs there, like Norway's *Ekofisk*, have been recording wave profiles continuously for as long as a decade. Therefore, Harald Krogstadt and friends decided to look for rogues in 100,000 samples of *Ekofisk* records. In 2008 they confirmed that the highest waves (rogues) are actually members of a normal population of random waves—although way out in the tail. A Raleigh distribution, modified to account for the cusps and deep troughs of big waves, fitted the data to a tee. Like the MAXWAVE team, these Norwegian researchers could not find any criterion, such as significant wave height, that might forecast a series of extreme waves. Therefore, forecasting rogue waves, just like trying to predict when a string of 10 heads might appear in repeated 10-toss games, remains a distant goal.

The latest statistics of rogue waves were compiled by Burkard Baschek (UCLA) and his colleague Jennifer Imai (California State University, Long Beach). In 2011 they examined buoy measurements on the U.S. west coast made over 81 years. They found over 7,100 rogues, with the tallest at 19 m high (62 ft). From these data they could estimate the likelihood of a rogue in the open sea (101 per year) and in coastal waters (63 per year). Along the world's major shipping lanes the chance is about 1% *per day* for a wave at least 11 m tall, but it need not qualify as a rogue. Of course, their estimates don't take into account special circumstances such as currents or a rough bottom topography, but they are interesting nevertheless.

We can use the following example to illustrate the statistics of random rogue waves. In 2005 an extreme wave in the open sea almost destroyed a ship. The 510-foot research ship *Explorer* was on route between Vancouver and Japan, carrying more than 680 students and 120 faculty and staff. On January 27 the ship was located 650 miles south of Alaska, plowing through heavy seas. Suddenly a 50-foot wave leaped out of the sea and flooded the bridge, knocking out the controls. Three of the four engines were disabled by the shock, and the crew could barely keep the bow of the ship headed into the high seas. It was at risk of turning broadside to the waves and capsizing. Fortunately, the crew did manage to use emergency steering methods to save the ship, and it limped back to port with only two injured crew members.

If we assume that this random 50-foot wave was a rogue and that the significant height was 25 feet, we can calculate that a wave of this height would arise only once every 3,000 waves. So if the average wave period was, say, 10 seconds, a random rogue would be expected every 8 hours. The *Explorer* was actually quite lucky not to be hit more than once.

What If Rogues Are More Frequent than Random?

Some scientists are not convinced that rogue waves are just the tails of a Rayleigh distribution. They think that special circumstances and special physics are required to produce at least some of them. They point to a probable connection between currents (such as the Agulhas) and rogues. Likewise, the high incidence of rogues in the shallow North Sea points to the influence of the sea floor topography. And they say that encountering three rogues in succession (the Three Sisters) would be impossibly improbable if rogues were truly random events. How is it possible that a rogue wave can arise from a calm sea, they ask. Finally, being told that a rogue is completely unpredictable and "just happens" leaves a scientist feeling quite unsatisfied.

For these reasons theorists began to explore other mechanisms that could explain rogue waves, with the hope of being able to predict them. Two classes of models have been thoroughly investigated in the past decade.

First are models that involve the *focusing* of wave energy. Focusing can occur in several ways: because of the different wave speeds within a group (dispersion), through the interaction of a group with a current, or by the crossing of wind-driven waves and swells (a cross-sea). All of these mechanisms are examples of constructive interference, a familiar "linear" process. Second, there are models that involve "nonlinear" processes, such as the intrinsic instability of a wave group or the interaction of a wave with itself. Somehow nonlinearity allows waves within a group to steal energy from their neighbors and grow to extreme heights. Below, I discuss both kinds of models, beginning with some linear models.

Rogues Caused by Wind Gusts

The simplest way to generate a single rogue wave in a random sea is to invoke a gust of wind. Imagine a mild or moderate sea made up of many weak sinusoidal waves of different wavelengths that are running in random directions. Now imagine that a gust of wind slaps the surface and launches a group of waves in the wind's direction. Such a group will have short waves in front and longer waves in back (as we saw in chapter 2). Within a few moments the longer and faster waves overtake the shorter and slower waves and overlap them. This process creates a high crest in the group.

If, in addition, the crest happens to coincide with a momentary crest in the random sea, a rogue wave can be formed. Remember, a rogue needs to be only twice the height of the significant wave height, not necessarily a very tall wave or a long-lived wave. This mechanism was described in great mathematical detail in a 2009 scholarly book by C. Kharif, E. Pelinovsky, and A. Slunyaev, *Rogue Waves in the Ocean*; and it was simulated by Efim Pelinovsky and his Russian colleagues in 2011. Kharif and colleagues claimed that this process would work even in a rough sea. They also estimated that such a rogue event would only last two minutes before the overlapping waves would again disperse.

This way of focusing wave energy, using the variation of wave speed with wavelength, has also been demonstrated in wave tank experiments. The scheme works well if the wave is restricted to travel in only one dimension, but other tank experiments show that a rogue is not generated if the wave is allowed to spread in several directions.

Rogues Caused by Currents

Another possible focusing mechanism involves strong currents such as the Agulhas or the Gulf Stream. The Agulhas Current is well known for generating big waves. Seasoned mariners tell of the "Cape rollers" that the southwestern gales drive northeast against this current. Swells 10 to 13 m high are not unusual. It is easy to picture a swell moving in the opposite direction to a current. When the swell meets the current, it slows down, shortening its wavelengths and rising abruptly in height. It might even be stopped dead and reflected backwards by the current.

The British ship *Waratah* probably met such conditions on her voyage from Melbourne, Australia, to London, with a stop at Durban, South Africa. She left Durban for Cape Town on July 26, 1909, with 211 passengers and crew aboard. Late in the day the wind picked up sharply, gusting to 50 knots (93 km/h) and raising waves 10 m high. The ship was seen battling the waves by observers on other ships, the last time 180 miles from Durban. The 465-foot *Waratah* never made it to Cape Town. She vanished without a trace. No one survived. It seems most likely that a rogue wave engulfed the ship, but without hard evidence it is impossible to say. Several attempts have been made to find the wreck but to no avail. Plans are afoot to use a submersible to search the bottom within a 50-mile radius of where she was last seen.

Let's return to the physics. Two British scientists—M. Longuet-Higgins and R. Stewart—were the first to describe the focusing of wave energy by a current. In 1961, long before rogue waves were recognized, they imagined a train of single-wavelength waves riding into a current. The current could either oppose or assist the wave. In their model, the volume of the current was steady, but its speed varied along its length, as would occur if the current were spreading out. They learned that if the current and the wave were moving in the same direction, the wave would lose energy to the current. But when the current opposed the wave, the wave would gain energy from the current. Basically, the current pushes back on the wave, exerting a steady force through a distance of several wavelengths. And that amounts to doing work on the wave. The wave rises in height as a consequence.

The effect depends on the relative speeds of current and wave. If, for example, the opposing current's speed was a quarter of the wave's initial speed, the wave's height would grow by a factor of 3. But the average speed of the Agulhas Current is only a few percent of the speed of wind-driven waves, so applying this analysis does not appear realistic.

In reality, however, a current like the Agulhas may have pockets of higher and lower speeds because of the rough bottom in this shallow area. Benjamin White and Bengt Fornberg examined the effects of such speed variations on a wave train in 1998. They postulated that the current has random fluctuations of speed in time and space, and they looked for any effects from these fluctuations on an oncoming wave.

In one of their numerical simulations, a plane wave front enters an opposing current that has pockets of fast and slow speeds. As different parts of the wave front encounter different water speeds, they are refracted, so that they bend in new directions. Each part of the front follows its own curving trajectory or ray. Where the rays cross, they interfere constructively to form what are called *caustics* (sunlight refracted through ripples onto the bottom of a shallow pond illustrates this effect), and the local amplitude rises. These are the places where a rogue wave might be formed. And as pockets of current speed change, the whole pattern of rays and rogues may change within a few moments.

This scenario works well only if the incident wave does not spread laterally (according to K. B. Dysthe). If it does spread, the caustics smear out, and the chance of forming a rogue is reduced. But because we're not absolutely sure of the frequency with which rogues appear, this reduction may not be fatal to the idea.

Nonlinear Models of Rogues

Much of the recent research on rogue waves invokes nonlinear processes. Only these, it is thought, can explain rogues that are too tall or too frequent to be explained by the simpler linear theories. Nonlinearity also is interesting in itself because of the unexpected effects it reveals.

We need to be clear about the meaning of nonlinearity. It refers to a change in a system that depends on the present state of the system. It usually involves a positive feedback loop, so that the system interacts with itself. As an example, consider compound interest. Part of the growth in your savings account comes from the interest you earned this year on the principle. That is the linear part of the growth. But you also gained from the interest on *last* year's interest; interest on interest. That is the nonlinear part, a kind of feedback loop. This nonlinear growth produces an exponential rise in wealth (if you don't spend the profit).

In a group of ocean waves, as we shall see, long waves can steal energy from shorter waves. That is a nonlinear process that wouldn't happen otherwise. In chapter 4, I briefly discussed the Stokes wave, which is the simplest example of a nonlinear wave. It interacts with itself to produce a sharper cusp and a flatter trough than a sinusoidal wave has. Waves in a swell resemble Stokes waves.

For a long time, it was thought that Stokes waves could propagate indefinitely without change of shape. Then in 1967, two British physicists discovered by accident that a train of Stokes waves can become unstable after traveling some distance in a wave tank. T. Brooke Benjamin was a brilliant English mathematical physicist and mathematician at Oxford University and a highly respected expert in fluid mechanics. Jim E. Feir was his student. This important discovery became known as the Benjamin-Feir Instability.

The two researchers intended to generate a train of waves with constant frequency and amplitude, but unexpectedly, their wave generator imposed a slow variation of amplitude along the length of the train. The wave train moved down the tank without incident until suddenly two new wave frequencies ("sidebands") appeared in the train. One frequency was slightly higher than the primary wave's frequency; the other was slightly lower. These sidebands grew exponentially in height at the expense of the primary wave, which eventually disintegrated. After eliminating all possible sources of equipment vibrations and imprecision, Benjamin and Feir determined that this nonlinear phenomenon was indeed real: the slight amplitude deviation was reinforced nonlinearly, leading to this sideband instability. Nobody had ever seen or expected the likes of this before.

Could this Benjamin-Feir instability produce a rogue wave? Many scientists thought so. To test the idea, they had to investigate the later stages of the instability, when the sidebands waves would grow to impressive heights. And to do that they employed a strange wave equation, the nonlinear Schrödinger equation. (We talked briefly about the origin of this equation in chapter 4.) Let's look at a recent example of its predictions. This numerical simulation was performed by Christian Kharif and Efim Pelinovsky in 2003.

In figure 8.2A we see the original wave. It has a constant wavelength and a slightly varying amplitude. After a certain time, the instability occurs and a single wave group is formed (fig. 8.2B). Its central wave has a crest-to-trough height almost *three times* the significant wave height. Here indeed is a rogue wave! A study of the instability showed that the wave group had extracted energy from the original wave in order to form the high central peak. In the later stages of the event (fig. 8.2C), both the original wave group and the tall central peak have disappeared and have been replaced by a chain of wave groups, each with a central peak twice as high as the average. Finally, assuming no dissipation in the model, the groups return their energy to the primary wave, which recovers almost completely (fig. 8.2D).

Other theorists have pursued variations of this attractive scenario. They discovered that if the whole chain of events repeats, then the wave groups pulsate periodically: they become "breathers." These early numerical simulations were highly artificial and were intended to show how well the Benjamin-Feir

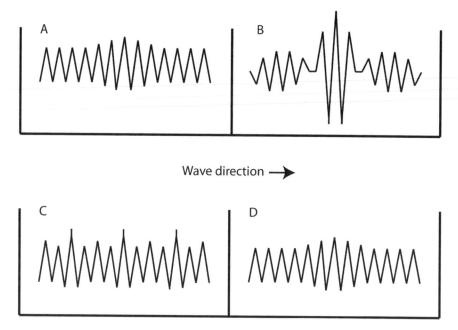

Wave direction ⟶

Fig. 8.2 The Benjamin Feir instability can generate a rogue at the center of a group of waves. The diagrams show the stages of the event: *A*, the original wave; *B*, single wave group formed after instability; *C*, subsequent chain of wave groups with a high central peak; *D*, return to original wave. (Drawn after C. Kharif and E. Pelinovsky, *European Journal of Mechanics, B: Fluids* 22 [2003]: 603.)

instability might work under ideal conditions. In particular, theorists wanted to learn how to model nonlinear waves with various exotic wave equations, such as the nonlinear Schrödinger, Dysthe, or Zakharov equations. The results were highly encouraging.

Several limiting factors for this scenario were discovered later, however. First, a wave train is unstable only if it spreads away from the wind direction by less than 35 degrees. In other words, a rogue forms only in a nearly unidirectional wave train. Secondly, the instability is less likely when the frequency spectrum of the initial wave train is broad. And finally, tank experiments suggest that even small amounts of dissipation—say, by wave breaking—may suppress the Benjamin-Feir instability entirely.

Perhaps the critical question is whether the Benjamin-Feir instability could arise at all in a random sea. And if it could, would a rogue wave break up before it reaches its maximum theoretical height? A. R. Osborne, M. Onerato, and M. Serio of the University of Torino tackled this difficult question in 2005. They asked under what conditions the Benjamin-Feir instability would increase the probability of a rogue wave in a random sea.

They began by calculating many examples of wave groups in a random sea. To do that, they selected one of the JONSWAP energy spectra and varied the phases of waves randomly. Then they calculated the evolution of each wave group for nine hours, using the nonlinear Schrödinger equation. That interval allowed different wavelengths to exchange energy. Finally, they calculated the distribution of wave heights in each group and averaged the results.

They learned that large-amplitude waves can appear from a random sea with a higher probability than the Rayleigh distribution forecasts, but only if the sea is very stormy—that is, only when the JONSWAP spectrum has a very high, narrow peak with most of the energy in a limited set of frequencies. Under those circumstances, very long wave groups are more likely, and the Benjamin-Feir instability also becomes more likely.

The Continuing Problem of Predicting Rogue Waves

An international conference on rogue waves is held every few years. One was held in Brest in November 2000, another in Hawaii in January 2005, and another in Edinburgh in December 2005. Clearly, the experts are learning a lot about nonlinear water waves. The theory and simulations are becoming very esoteric, although tank experiments are helping to guide the theory. Some scientists are combining linear and nonlinear effects, such as the instabilities caused by crossing seas. Others are showing that even slight dissipation or damping of waves can mitigate the instabilities and can therefore prevent the generation of rogue waves. So the search and research goes on.

But the prospects of predicting rogue waves, or their precursors, seem as remote as ever. And yet that is what the operators of oil rigs and captains of container ships most desire. They now know what locations are best to avoid, if they can. But it's difficult to avoid a rogue in the open sea, far from a coast. Perhaps a major experiment at sea, like JONSWAP, is needed to clarify the science.

CHAPTER NINE

Tsunamis

The 2004 Indian Ocean Tsunami

On December 26, 2004, at 7:58 a.m., the third largest earthquake ever recorded by seismographs occurred 160 km off the northeast coast of the island of Sumatra. The earthquake was rated between magnitude 9.1 and 9.3 and broke the record for duration: 8 to 10 minutes. Seismologists determined that the epicenter lay at a depth of 30 km below the surface of the ocean.

This violent earthquake triggered a tsunami, a high-speed water wave of terrifying power that spread in all directions from the epicenter. The eastern side of the wave roared across the short distance to the coast of Sumatra as a wall of water 10 m high. The western side ran out into the Indian Ocean at a speed of about 720 km/h. It reached Sri Lanka and eastern India a couple of hours later, then swirled around to hit Phuket in Thailand. Finally, 7 hours later it swept across the Horn of Africa.

The coastal wave swept inland in Sumatra, destroying everything in its path. The four main towns in the province of Aceh, each home to as many as 12,000 people, disappeared within minutes. Of 28 villages along the coast, only 4 survived. At the north-facing city of Banda Aceh, the oncoming wall of water was 11 m high. At the west-facing villages of Meulaboh and Tapaktuan, the wall towered an astounding 30 m (98 ft) high. In fact, the wall measured between 15 and 30 m high for 100 km along the coast, from Kreung Sabe to the northwest tip of Sumatra. The water poured inland for 10 km and more, bulldozing houses, cars, and trees, tearing up roads and bridges, pulling down power lines.

A TV reporter broke down in tears as she saw the devastation at Banda Aceh. "The city's like a huge dumpster," she said. "Dead bodies still scattered everywhere, debris blocking the streets with possibly victims of the tsunami in those piles. Help hasn't arrived yet even there in the capital of Aceh. People who are hungry tried to break in one grocery store only to find food covered in mud and water and bodies of people who couldn't escape at the time the tsunami hit" (BBC News, Jan. 6, 2005).

In Phuket, Thailand, 2,300 km from the epicenter, a 14-year-old girl gave the following eyewitness report, printed in the *Guardian* (Dec. 22, 2006):

Dad is looking out to sea in a strange way. Mum and I look up and see the water disappear, leaving all the fish on the sand. We see children running out to help the fish back into the water, so that they do not die. Dad wants me to fetch the camera from the hotel so that we can film the water disappearing. I am too lazy. Dad gets up to fetch it himself, but first he and Mum have a little argument. Dad thinks that the water is drawing out. Mum and I shriek, "It's coming in."

"Calm down—of course it isn't coming in," says Dad, on his way to the hotel. I have not seen him since.

Mum and I see the wave. We take our stuff and run. Mum runs away ahead of me. I hear her voice: "For goodness sake run, Charlotte! Whatever happens I will always love you." I have not seen her since.

She disappears without bothering to check whether I am behind her. I run in panic, upwards, as far as I can. Get to a flight of steps where there is chaos. A small child is standing by the steps crying. The mother has left the child alone.

I am holding tightly on to the stair rail when the wave roars in over the whole of Khao Lak. I feel the wave rolling over me and pulling away the rail. I go with the wave out to sea and in again, several times. Under the surface, I swallow gulps of salty water when I try to get air. I will not survive if I do not come up to the surface. In the end I can take deep breaths. With my eyes closed. I am hanging in something, a tree? The roof of a house? The thing I am hanging on snaps and I am pulled out to sea again, out and in. After perhaps seven minutes I open my eyes. I have landed up by the hotel and see masses of people lying there, blood everywhere.

The girl's brother was watching a film in the hotel when the wave hit. Here is his report:

I lay down on the bed to watch "The Day After Tomorrow," a film about an enormous wave. Quite weird it was that film. Mum, Dad and Lotti went down to the beach.

The electricity is cut off. There is a sudden sound like thunder and heavy rain. I look out of the window—and see a wave that must be 15 m high. I see bungalows being swept away, and cars, and people lying bleeding, and I understand nothing. I grab the room key and want to run down to the beach, but all the paths and steps have gone. Get very worried for Mum, Dad and Lotti. See people floating in the water and start crying.

The children never saw their parents alive again.

In Sri Lanka approximately 35,000 people died; in the eastern and southern coasts of India, the toll was 9,000; in the Andaman and Nicobar Islands, 5,000 died. In Thailand, a three-story wave hit hotels and villages, killing 5,000. In Somalia, 5,570 km from the epicenter, entire villages were swamped and 300 died. Many aftershocks were recorded at the epicenter for several days. One of them had a magnitude of 8.7 and could be considered a "triggered earthquake." None of these aftershocks produced a severe tsunami, however.

When the dead and missing were counted, weeks after the disaster, the toll had reached a staggering 230,000, spread over 14 countries. The great majority (137,000 dead and 37,000 missing) were in Indonesia, where the wave hit hardest and without warning. In addition, over half a million people were displaced, and millions were left homeless. A massive international rescue effort was mobilized to help the survivors, but eight years after the event, the people of Aceh province are still trying to resume their former lives.

The Origins of Tsunamis

Fortunately, tsunamis as powerful as this 2004 event are rare. They are generated primarily by submarine earthquakes but can also be caused by undersea landslides and volcanoes. Submarine landslides can also generate huge tsunamis. While tsunamis associated with earthquakes are limited in height by the magnitude of the quake, a landslide tsunami is limited only by the vertical distance of the slide. An underwater slide can begin on the continental slope at a depth of a hundred meters and end at a depth of several thousand meters. The gravitational energy released in such a slide can create a monster tsunami.

The most common cause of a tsunami produced by an erupting volcano is a pyroclastic flow of superheated gases and rocks that has erupted from a volcanic vent and either flows into the sea at very high speeds or bursts out from a submerged vent. In either case, the pyroclastic material displaces huge volumes of water, which bulge up into a gigantic wave.

Tsunamis caused by landslides and volcanoes can often be far larger than earthquake-caused tsunamis. Tsunamis can also be triggered by calving glaciers and meteorite impacts, although these are generally smaller than landslide-caused tsunamis.

In the Pacific Ocean the sources of the earthquakes (and of many volcanic eruptions) lie in a narrow band around the perimeter, the so-called Ring of Fire. It touches the whole western coast of North and South America, the Aleutian Islands, the east coast of Asia, Indonesia, and the chain of islands that terminates at New Zealand. This band marks the boundaries between colliding tectonic plates.

As you may recall, the crust of the earth is divided into approximately 18 tectonic plates of varying sizes, each about 100 km thick. The plates move horizontally in different directions, a few centimeters per year. They are driven by slowly turning convection cells in the underlying molten mantle of the earth.

Where two plates meet, they may collide head-on and raise a ridge of rock that forms a mountain chain. (The Himalayas, for example, are formed by the collision of the Indian and Eurasian plates.) Two plates may diverge so that hot lava seeps up through the crack between them, quickly cooling to form new plate material. (The mid-ocean ridges that wind around the globe are formed this way.) Alternatively, plates may slide past each other, as happens on the San Andreas fault, along the west coast of North America. Finally, one plate may submerge beneath the other, a process called subduction. Subduction is the main type of plate interaction for the Ring of Fire.

In subduction, the plate motions need not be smooth. The plates may jam so tightly that an enormous stress builds up in the rock at the junction. At some critical point, the logjam breaks, and one plate shifts violently and thrusts suddenly beneath the other.

That is what happened in the Indonesian earthquake. The Indian tectonic plate (part of the Indo-Australian plate) thrust under the Burma tectonic plate (part of the Eurasian plate) at the Sunda trench, which stretches along the western coast of Sumatra. A section of this Sunda trench, assumed by many geologists to be dormant, ruptured violently along a length of 1,600 km. It slipped sideways about 15 m at a speed of 2 km/s, like a crack opening on a frozen lake. The seafloor on the overriding Burma plate uplifted seaward (toward the trench) and downward toward the coast, by about 4 m in just seconds.

This vertical movement caused the tsunami. The whole water column, from the bottom to the surface, and over an area of 100,000 km^2, was lifted.

The water rose into a great dome within a few seconds. Geologists estimate that the dome contained $30\,km^3$ of water. When the dome collapsed under its own gravity, it pressed against the water to the sides and launched a water wave that carried away the potential energy of the dome. This was the tsunami.

As a result of the 4-m vertical motion, an N-shaped water wave was created. The wave immediately split into two N-shaped waves, traveling in opposite directions. One wave headed west to the Indian Ocean with a leading crest. The other wave headed east toward the coast of Sumatra with a leading trough. Each of these two tsunamis propagated as a shallow-water wave, whose speed is determined by the local depth of the bottom.

The trough running toward shore encountered shallower water, and therefore slowed down. Its wavelength decreased as well. To conserve its energy, the wave's amplitude increased, and a towering crest was created. In contrast, the seagoing wave was crossing the 4-km-deep Indian Ocean at a height of a few meters at most, and with a wavelength of tens of kilometers. It would have been hardly noticeable to a ship at sea. It wasn't until it reached Sri Lanka and the Indian coast that the crest also rose to terrifying heights, causing tremendous loss of life a couple of hours after the earthquake. Some of the tsunami's energy reached South Africa and even spread into the Atlantic and Pacific Oceans. Tiny tsunamis were registered there, possibly funneled by the mid-oceanic ridges, to reach South America and even Vancouver in Canada.

When the crest of the tsunami arrives at a shore, it builds to a great height. Most tsunamis run up a shore like a rapidly rising tide (which is why they are sometimes called tidal waves), but a few are affected enough by the undersea reefs and the slope of the beach to become breaking waves. The Indonesian Ocean tsunami was one of these few; it arrived as a breaking wave, similar to the surf that one commonly sees at a beach. Much of the damage the tsunami inflicted was caused by the swirling currents, loaded with the debris that it generated.

Earthquakes and Tsunamis

Most tsunamis (58%) are produced by earthquakes where tectonic plates collide. But not all submarine earthquakes produce tsunamis. Russian geologist Victor Gusiakov has looked into the matter. The main requirement is that the earthquake vertically displace the overlying water; slip-strike earthquakes (those sliding sideways) do not create tsunamis. In addition, as you might expect, the greater the magnitude of a quake, the higher the chance that it will produce a tsunami. Gusiakov showed that a quake as large as magnitude 6.7–6.9 has only a 12% chance of launching a tsunami, while only half of the quakes with magnitude 7.0–7.4 produce a tsunami. Earthquakes above magnitude 8 all produce tsunamis, generating 40% of the large transoceanic tsunamis, although not all have such devastating impacts as the 2004 Indian Ocean tsunami.

Whether an earthquake will generate a tsunami and how destructive that tsunami might be are difficult to predict. For one thing, it is not immediately clear how much vertical displacement of the water might have occurred. Secondly, the magnitude of a quake has very little relation to the energy of the tsunami it may produce because the total earthquake energy is related to the total ruptured area, not just to the maximum intensity at the epicenter. For example, the tsunami that a magnitude 7 quake produces can vary in energy by a factor of a million or more, depending upon the type and depth of the earthquake and the area of the rupture.

As a case in point, the earthquake of May 22, 1960, at Valdivia, Chile, is still the largest earthquake ever recorded, with a magnitude of 9.5. It generated a tsunami that battered the coast of Chile with waves as high as 25 m. Houses were pushed 3 km inland by the water. The tsunami crossed the Pacific and killed 61 people in Hilo, Hawaii; 138 in Japan; and 32 in the Philippines. It completely destroyed the port of Hilo and caused extensive damage on the west coast of the United States. Waves as high as 10 m struck Honshu, Japan, 22 hours after the quake. But despite the 9.5 magnitude of this earthquake, the resulting tsunami had significantly less energy and caused only a fraction of the total destruction and death resulting from the (slightly) smaller Indian Ocean earthquake.

The degree of destruction and death caused by a tsunami also depends on the nature of the coastline, the steepness of the shore, the nature of the seabed, the proximity to the epicenter—and the awareness of the population to the potential danger from tsunamis. For instance, after the 2004 Indian Ocean earthquake, the aboriginal population of the Andaman Islands recognized the signs of the approaching tsunami and fled to safety, while their more "civilized" neighbors perished in the waves. Even 10-year-old Tilly Smith, playing on the beach at Phuket, Thailand, recognized the receding water as the onset of the tsunami and was able to get her family and others on the beach to safety.

A Chain of Catastrophes: The March 2011 Disaster in Japan

Sometimes the aftereffects of an earthquake and tsunami are just as devastating as the original event. This was the case with the magnitude 9.0 quake that struck the east coast of Japan at 2:46 p.m. local time on Saturday, March 11, 2011. It was the largest quake ever recorded in the long history of Japanese earthquakes.

The epicenter lay 100 km off the coast, at a depth of 6 km. There, the Pacific tectonic plate had been driving down under the Eurasian plate for centuries. Stress in the rocks kept building and building. On March 11, the fault snapped. The seabed jolted upward 5–8 m along the 480-km fault line, raising a gigantic volume of water in a dome a few meters high. Moments later, as the dome collapsed, it generated an immense tsunami that contained nearly double the energy of the Indian Ocean tsunami—enough energy to power the city of Los Angeles for a year.

Because of its huge wavelength, the tsunami propagated as a shallow water wave, whose speed was determined by the depth of the seabed. Part of the wave moved inexorably toward the shore, averaging over 100 km/h, and part of the wave raced out across the Pacific toward Hawaii and the west coast of North America, at closer to 800 km/h.

As the tsunami wave got closer to the Japanese coast, the front of the wave slowed even more because of the increasingly shallow shore, and the water at the rear piled up into a towering wave. About 30 minutes after the earthquake, the first wave hit the city of Ofunato as a wall of water probably 10–20 m (34–68 ft) high (analyses of these heights vary tremendously), sweeping everything before it. Down the coast, the city of Sendai was inundated next. The land around the city was dead flat, so that nothing stood in the way of the raging wave (fig. 9.1). A roiling mass of water and debris flooded inland as much as 7 km. The surge climbed hills higher than 40 m.

Somewhat farther away from the earthquake's epicenter, the citizens of Miyako had a 60-minute warning to seek high ground. Years before, they had built a 10-m-high seawall to protect their city from just such calamities. Therefore, assuming they were well protected, about 40% of the people ignored the warning. But to their shock and horror, a 40-m tsunami roared over the wall easily, surged across the coastal road, and drowned the city (fig. 9.2). Boats, cars, trees, and small buildings were washed away like toys. Half of those who

Fig. 9.1 An aerial view of the Sendai airport on the coast of Japan on March 11, 2011. (Wikipedia.)

disregarded the warning perished in the powerful churning waters. Later, after the danger had passed, geologists learned that the earthquake had dropped the shoreline by half a meter, allowing the tsunami even easier access to the city.

More than 19,000 Japanese were counted as missing or confirmed dead in the catastrophe. They were either swept out to sea or drowned. Farther to the east, the tsunami also caused serious damage in Hawaii and in some places along the western coast of the United States. Thousands of birds nesting on Midway's low-lying islands were swept to their deaths. But except for one unlucky California man (attempting to photograph the tsunami), no humans outside of Japan lost their lives.

However, the monster tsunami created another serious threat to the population: the inundation of the Fukushima Daiichi nuclear power station. The plant had six reactor units, two of which were in cold shutdown for routine maintenance, while a third one had had its fuel rods removed in preparation for refueling. When the earthquake struck, the remaining three reactors automatically scrammed (shut down) exactly as they were supposed to. Power to the Fukushima plant was also severed as a result of the earthquake. Emergency generators clicked on immediately to maintain the flow of cooling water through the hot reactor cores.

Fig. 9.2 A view of Miyako on March 11, 2011. (Wikipedia.)

But then 50 minutes later, the 15-m tsunami arrived, easily topping the seawall and completely flooding the underground chamber where the generators were located. Unfortunately, this chamber had not been sealed, nor had the generators been moved to higher locations, as recommended by Japanese nuclear safety regulations. The generators drowned, the pumps that cooled the reactors shut down, and as the cooling water started to boil away and expose the nuclear core, the reactors heated up rapidly. Nothing was happening as it was supposed to anymore—the plant engineers were operating in unknown territory.

Officials realized that without cooling water, the reactor cores could melt down and cause a nuclear disaster comparable to Chernobyl. As a last resort, knowing it would destroy the reactors, they ordered seawater to be pumped onto the reactors in a desperate attempt to provide some cooling. First, they used fire truck pumps, but they eventually got three seawater pumps repaired from the tsunami damage, while crucial electrical equipment to run them was found in good shape on higher ground, as the regulations required. Nonetheless, the pressures within the reactors continued to build, and the hydrogen gases created by chemical reactions with the reactor material eventually caused explosions at all three reactors over the next couple of days.

Additional radioactive gases were released to reduce the pressure. But it finally became clear that all three reactors had experienced partial meltdowns. The fuel rods had become exposed to air and had heated up so much that they melted the bottom of the reactors and fell to the concrete floor below. The high levels of radioactivity measured at the plant hampered actions by the plant operators. More than 200,000 residents who were within 20 km of the plant were hurriedly evacuated from the vicinity.

Not even reactor 4 escaped the disaster, although it was shut down. On Tuesday, three days after the tsunami, a fire broke out on the roof of this reactor, where its spent fuel rods were cooling, melting some of the rods and releasing even more radioactivity into the atmosphere.

The key to the emergency was the lack of power to cool the reactors. In the following weeks, workers risked their lives to build a power line into the plant. Eventually the power line was completed, power and cooling was restored, and the immediate threat of total nuclear meltdown was removed. But the radiation level was a thousand times normal within the plant and eight times normal in the surrounding countryside. It had poisoned hundreds of square kilometers of land, crops, and buildings, while several hundred thousand residents of the area lost their homes with no hope of ever returning. Many had lost everything they owned—their homes, their clothing, their pets. How does one cope in such a situation? How does the government grapple with the most immediate problems?

The government mobilized 100,000 troops to help in the relief effort. Nations around the world contributed to the relief of the stricken Japanese. Following the disaster, the Japanese government announced its decision to close down its nuclear power plants, in recognition of the public's revulsion and fear. Recently, some plants are being restarted because Japan lacks viable alternative energy sources.

As with so many accidents, it was the unforeseen sequence of events that caused the worst of the devastation. Although the catastrophe was triggered by a natural disaster, one failure after another created havoc. Electricity failed, cell phones failed, trains stopped running, roads were impassible, and the radiation limited access to critical areas. However, human errors and misjudgments exacerbated the situation. One of the key misjudgments was the complacency of so many people that anti-tsunami seawalls could eliminate the risk from these hellish waves. The enormously energetic surge just washed right over them, knocking them down like children's blocks. Those living near the ocean need to learn from their ancestors, who planted stone markers on hills showing where tsunamis had reached in their day—some still many meters above the reach of the March 2011 tsunami.

Historic Deadly Tsunamis

Japan has probably suffered more tsunamis than any other nation, averaging one about every 6 years. The first recorded tsunami occurred at Hakuho on November 24, 684 CE. The preceding earthquake has been estimated at a magnitude of 8.4. There followed a sequence of notable earthquakes and tsunamis, in 869 and 887. In 1293 a 7.1 magnitude earthquake and a tsunami struck the city of Kamakura, killing 23,000. Disaster struck again in 1361 with an 8.4 magnitude quake and again in 1498 with a 7.5 magnitude quake that left 30,000 dead. On February 3, 1605, an 8.1 magnitude quake triggered tsunamis with heights of 6–8 m and caused more than 5,000 deaths by drowning. On October 28, 1707, an 8.4 quake triggered a tsunami with a reported height of 10 m. Nearly 50,000 were killed. But none of these could compare with the Indonesian event of 2004, with a death toll of 230,000.

Deadly tsunamis are not limited to the Pacific Ocean. The Atlantic Ocean and Mediterranean Sea are also famous for such disasters. The earliest geological record is of a landslide-caused tsunami in the Norwegian Sea about 6100 BCE; scientists estimate that it deposited material from the sea at least 80 km inland.

Around 1600 BCE a great volcano erupted on the island of Santorini in the Aegean Sea. The whole top of the island disappeared, leaving a semicircular caldera that now forms a tranquil bay. The eruption occurred at the height of the Minoan civilization, based on the island of Crete, which lies some 100 km from Santorini. Recent research suggests that a powerful tsunami inundated the coast of Crete, killing thousands and leading to the decline of the Minoans. In 373 BCE, the Greek city of Helike, 2 km from the sea, was struck by a tsunami and remained permanently submerged. It may have been the source of the legends of Atlantis.

On July 21, 365, a powerful earthquake triggered a monster tsunami, reportedly 30 m high, that devastated the ancient city of Alexandria. Thousands were killed when, after the water suddenly retreated from the shore, they raced to capture the flapping fish that had been stranded in the slime.

Just as suddenly, they were overtaken by the tsunami as the surge swept back in, scattering ships and bodies far inland. The disaster was described in vivid detail by a Roman historian, Ammianus Marcellinus.

One of the most famous earthquakes and tsunamis occurred on November 1, 1755, off the coast of Portugal. The capital city of Lisbon first was devastated by the earthquake and by the fires that broke out afterward. Tens of thousands fled to the waterfront to escape the fires. Forty minutes later, they were awestruck to see the sea recede, stranding ships in the harbor. Then a great wave, estimated at 15 m in height, broke over the people, and flooded the shore for kilometers inland. The total loss of life from all causes could have been as high as 100,000, making this one of the deadliest earthquakes in history and certainly the greatest catastrophe of the century in Europe.

In August 1883 a series of spectacular volcanic eruptions occurred on the island of Krakatoa (Krakatau), which lies between Sumatra and Java in the Sunda Strait. Four explosions ejected several cubic kilometers of ash into the atmosphere and were heard 3,500 km away in Perth, Australia. Tsunami waves as high as 30 m were generated by the pyroclastic flows of superheated gas and rocks that fell into the sea. The town of Merek, on the northwest tip of Java, was obliterated by a tsunami reportedly 46 m high. The waves swamped large areas of the coast of Sumatra and traveled as far as South Africa.

The eruption of Krakatoa also had other major impacts on the earth's climate. The ejection of massive amounts of sulfur dioxide into the atmosphere led to the formation of sulfuric acid, which then promoted the formation of dense cirrus clouds. High-level winds carried the cirrus around the globe. These white clouds reflected sunlight more efficiently and caused a global drop in temperature of about 1.2°C. The cooling effect lasted for several years.

One of the largest recorded tsunamis was caused by an underwater landslide at Unimak Island, the easternmost island in the Aleutian chain, on April 1, 1946. The slide, triggered by a magnitude 8.6 earthquake, contained about 200 km^3 of sediments. It started at a depth of 150 m and ended on the Aleutian Terrace at a depth of 6,000 m. Several tsunamis were generated, ranging in height from 15 m to an astounding 43 m (141 ft). A 35-m wave destroyed the lighthouse at Scotch Cap on Unimak and killed all five lighthouse keepers. Another wave crossed the Pacific, reached Hilo, Hawaii, after five hours and caused 159 casualties, as well as tens of millions of dollars of property damage. This tsunami is sometimes referred to as the April Fools Day tsunami because many people in Hawaii thought the tsunami warning was just a prank. It did result in the establishment of the Pacific Tsunami Warning Center in 1949.

The highest tsunami wave ever recorded occurred as a result of a catastrophic landslide into a bay. On July 9, 1958, an 8.3 magnitude earthquake caused 30 million cubic meters of rock to fall into the inlet of Lituya Bay in southeastern Alaska. Focused by the narrow width of the inlet, the wave rose to over 524 m (1,720 ft), stripping all trees below that height from the borders of the bay. The wave killed only three people on an island in the bay and two men in a fishing boat.

A similar event occurred in 1980. Mount St. Helens, an active volcano in Washington State, erupted explosively on May 18. The top 400 m of the mountain disappeared and was replaced by a mile-wide horseshoe-shaped crater. The eruption also triggered a massive pyroclastic avalanche, which fell into Spirit Lake, below the mountain. The resulting wave was 260 m high, as judged from the high-water mark on the opposite hillside.

Warning Systems

Most tsunamis are caused by submarine earthquakes, and to this date, no one is able to predict with any certainty when an earthquake and its associated tsunami will occur. However, an earthquake can be detected almost immediately by sensitive seismographs because seismic waves travel around the world at about 14,000 km/h. In contrast, a tsunami travels at less than 1,000 km/h. Therefore, there can be a delay of several hours between the earthquake and the arrival of a tsunami at some distance. This delay could allow a population under threat to be warned in time to escape to high ground. But such a potential saving of lives requires constant monitoring with a widespread system of sensors.

After the 1946 Aleutian Islands tsunami destroyed the waterfront of Hilo and killed 165 people, the U.S. government organized the Pacific Tsunami Warning System. It consisted of the existing seismograph at the Honolulu Geomagnetic Observatory and a new network of sirens that was spread throughout the Hawaiian Islands and coupled by telephone to the observatory.

Then in 1960, following the huge Chilean earthquake and its disastrous tsunami, 26 nations around the Pacific Ocean agreed to join in establishing the Tsunami Warning System. This was accomplished in 1965 under the auspices of UNESCO. Two centers were established to collect and analyze earthquake signals, determine whether a tsunami is impending, estimate the path of the tsunami, and if necessary, issue appropriate warnings. The U.S. National Oceanic and Atmospheric Agency (NOAA) operates the two regional centers. One is at Ewa Beach, Hawaii; the other is at Palmer, Alaska. The characteristics of an earthquake are recorded by seismographs at the two centers, by the U.S. Geological Survey's National Earthquake Information Center, and by international stations.

The other leg of the Tsunami Warning System is a fleet of instrumented buoys that signal the passage of a tsunami. Beginning in 2000, NOAA distributed seven buoys around the perimeter of the Pacific Ocean. These buoys were the first in the Deep-Ocean Assessment and Reporting of Tsunamis (DART) system. With additions by the DART partners, the Pacific system has expanded to more than 50 buoys, with a few sprinkled in the Atlantic and Indian Oceans. Ideally, one would want more buoys out in the center of the Pacific to be able to monitor a tsunami as it sweeps across the ocean.

The idea behind the buoys is that the average water pressure near the bottom of the ocean is hardly affected by the rapidly fluctuating, wind-driven waves at the surface. The bottom pressure merely reflects the height of the water column. But any long, slow change in the average pressure, beyond a set threshold, can indicate the passage of a tsunami.

Each DART II buoy (the improved version of DART I) is anchored to the ocean bottom and is equipped with a bidirectional communication system. A pressure gauge and a battery-powered electronic package lie on the bottom near the buoy. Water pressure is recorded continuously and averaged in 15-minute blocks by the electronic unit. The pressure gauge is extraordinarily sensitive; it can resolve a 1-mm change in the 4-km height of the water column. The data averages are transmitted to the buoy at the surface by means of an acoustic link and then sent on to one of NOAA's Geostationary Operational Environmental satellites and to ground stations. From there a warning can be distributed to various national centers. For example, during the February 27, 2010, tsunami at Chile, the average water column rose by 20 cm over a period of 20 minutes as the crest arrived at a buoy and then decreased as the trough passed by. A warning was issued shortly afterwards.

The seismic and water pressure observations are used to sort through a computerized generation and propagation database and to select the most likely scenarios. This step requires only a few minutes to complete. Then these deep-water estimates are used as initial conditions to compute the propagation of the tsunami and its height when it arrives at selected coastal areas.

This step can be completed in about 10 minutes. At especially vulnerable sites the shape of the bottom and slope of the shelf are taken into account. In recent years a statistical or probability technique has been used to estimate the range of hazards that a chosen coast can experience. Warnings are sent out by the member state and local governments, who spread the word by radio and television. NOAA's Weather Radio System, which is directly available to the public, is also used to warn the public.

The system performed very well for the United States during the March 2011 tsunami in Japan. The wave reached Kona, Hawaii, after 7 hours with a height of about a meter. It caused significant flooding but no casualties. It touched the yacht harbor at Santa Cruz, California, in 14 hours, causing minor damage to boats and the dock.

Prediction techniques are constantly improving, thanks to Vasily Titov and his team at NOAA's Center for Tsunami Research in Seattle, Washington. Not only can the height and time of arrival be forecast, but also the impact on a coastal zone. Titov was especially proud of the performance of his computer model during the April 11, 2012, tsunami off the western coast of Sumatra. This wave was generated by a magnitude 8.3 earthquake 437 km southwest of Banda Aceh, a town made famous in the great tsunami of 2004. Although the earthquake caused widespread panic in Banda Aceh, the model accurately predicted that a tsunami was not a serious threat to the town. In contrast, the model severely underestimated the tsunami's height at Hanimadhoo in the Maldive Islands. Tidal gauges there showed that the arrival time was accurately forecast but that the model failed to predict the 40-cm height of the wave. Fortunately, such a weak wave caused relatively little damage.

The terrible loss of life following the 2004 Indonesian earthquake and tsunami sparked an effort to set up a tsunami warning system for the Indian Ocean similar to that for the Pacific Ocean. UNESCO organized the system, which began operation in June 2006. It consists of 26 seismic centers, linked by satellite communications. The system performed adequately just one month later, on July 17, 2006, when a 7.7 magnitude quake and a 3-m tsunami occurred off the southwest coast of Java. The Pacific Tsunami Warning Center issued a tsunami warning about 12 minutes after the earthquake, but it was not always relayed in time to all populations in harm's way, since there was concern about "not alarming the people unnecessarily." Consequently, about 700 people died in Java because of the 2-m waves that swept inland at least

200 m. No system can protect people who live close to an epicenter or who are not warned in a timely manner.

A New Idea

As we have seen, the existing tsunami detection system depends on a fleet of DART buoys along the earthquake-prone coasts of the Pacific. There is now a proposal for a new detection technique that might be able to identify tsunamis on the high seas. It would involve the Global Positioning System (GPS) and receivers on commercial ships. Several scientists have suggested the idea independently. Here is the story of how one of them, John Foster of the University of Hawaii, got the idea.

In 2010 Foster was a member of the crew of the oceanographic research ship *Kilo Moana*. On February 27, while the ship was in transit from Hawaii to Guam, a magnitude 8.8 earthquake occurred at Maule, Chile. Foster and his colleagues were able to analyze the continuous GPS recordings of the height of their ship above mean sea level. That allowed them to detect the Chilean tsunami, although the wave was only 9 cm high. Foster was astounded. Nobody had expected to be able to detect a tsunami at sea because its height is usually so small. But the GPS location of the ship in three dimensions was sufficiently precise, with a resolution of just a few centimeters, to allow detection. The long wavelength of a tsunami (tens or hundreds of kilometers) is the key to distinguishing it from wind-driven surface waves.

Foster visualizes a network of shipborne GPS recorders linked by satellites that could detect tsunamis anywhere over the Pacific. As of March 2012 Foster and his colleagues were planning to test the feasibility of the scheme by linking two ships. Their plan was to stream GPS data continuously to determine the height of the sea in real time with sufficient accuracy to detect a tsunami. We shall have to see whether they are successful.

Tsunamis continue to inspire awe, fear, and morbid fascination in the general public. The opportunity of watching one in action on television or streaming video is irresistible. But for the people vulnerable to them, better warning systems—and the education to heed the warnings—are crucial to minimizing the devastation caused by these nightmarish water monsters.

CHAPTER TEN

Internal Waves and El Niño

Thus far we have discussed ocean waves that are easily visible at the surface. But there is another kind of wave that, although predicted 70 years ago, remained undetected until recently. These "internal waves" propagate on the thermocline, the narrow boundary between the warm surface layer of the ocean and the cold deeper waters. They cause barely a ripple at the surface, a few centimeters at most. And yet they affect weather and climate conditions around the world.

Two main types of these so-called planetary waves—Kelvin waves and Rossby waves—were discovered by our old friend, Lord Kelvin in 1879, and explained mathematically by Carl-Gustaf Rossby, a Swedish meteorologist, in 1939. These ocean waves participate in a dramatic global climate phenomenon, the El Niño–La Niña cycle. We'll begin our story with it.

El Niño

One of the most productive fishing grounds in the world lies off the coasts of Peru and Ecuador. Fish are abundant there because of a favorable weather pattern. During normal years, the strong trade winds from the southeast blow the warm surface waters far out to sea. Cold water wells up from the deep to replace the warm water. The cold water brings plankton and algae up from the depths, ensuring an abundant supply of food for small fish. As a result the coastal waters are filled with dense clouds of sardines and anchovies.

For many years Peruvian fisherman have made a good living from the sea. They could depend on a steady, constantly renewing crop of fish. But every five years or so, generally around Christmas, the trade winds fade for a few months. The offshore waters turn warm, the food supply shrinks, the fish migrate north or south, and the Peruvian fishing industry crashes. The fishermen call these episodes "El Niño," meaning "the Christ Child." They last for about a year and cause misery and deprivation among the fishermen. But then the trade winds start blowing, the cold water wells up at the Peruvian coast again, and the fish return.

El Niño events often have disastrous consequences far away from the Peruvian coast. In the severe El Niño episode of 1997–98 (the strongest in 100 years), significant details were learned about both ocean and atmospheric conditions from temperature data collected by instrumented buoys and meteorological stations.

Throughout most of 1997 La Niña was in force. As is typical of that situation, early in that year a low-pressure zone sat over Indonesia while violent thunderstorms and heavy rain battered eastern Asia. A pool of warm water (30°C) extended to a depth of about 200 m on the equator of the western Pacific. The warm sea caused moist air to rise by convection. The air traveled eastward and descended on the opposite side of the South Pacific, near Easter Island. The descending air produced a high-pressure zone, causing drought in Peru and Chile but also causing the trade winds to blow the warm waters away from the coast. A cool layer (22°C) of upwelling water only 50 m deep lay off the eastern Pacific along the equatorial coast of South America, providing the Peruvian fishermen with excellent fishing.

But by November 1997 the southeast and northeast trade winds had weakened considerably and the temperature distribution had reversed. Now, an El Niño event was in full force, with warm waters replacing the cooler waters and extending to a depth of about 70 m on the eastern equator of the Pacific (fig. 10.1).

In the western Pacific, Australia, Indonesia, and the Philippines suffered devastating drought, while in the United States, the change in weather was dramatic. The Northwest and the Midwest were warmer and drier than normal with much less snow. Large waves during the El Niño winter months eroded 41 of 47 beaches on Washington State's coast. High storm waves stripped sand from the shore, leaving many beaches narrower and steeper.

In the Northeast, warmer than usual temperatures minimized the snowfall. It may be that Quebec's terribly destructive ice storm that winter was exacerbated by El Niño's warming effects, which turned the normally dry snow into wet mush that froze overnight on trees and power lines, pulling them down in a tangle of branches and wires. At the same time both California

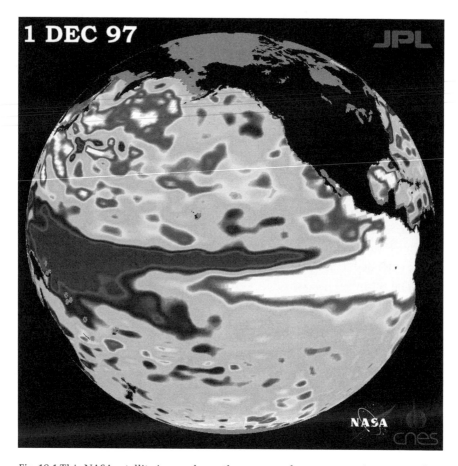

Fig. 10.1 This NASA satellite image shows the warm surface water moving eastward to the South American coast during the severe El Niño episode in 1997–98.

and the southeastern states experienced very heavy rain. Altogether, the damage to industry, crops, and property cost billions of dollars.

By March 1998, the trade winds had recovered and began to push the warm surface water back across the Pacific, bringing back La Niña.

Peruvian fishermen would like to think that La Niña periods are "normal" and El Niño events are the anomalies. But in fact both El Niños and La Niñas are part of an irregular cycle of atmospheric pressures called the Southern Oscillation. The Southern Oscillation is not strictly periodic. El Niños occur at intervals of 3–7 years, and an El Niño may be followed by two or more La

Niñas. It is an infuriatingly difficult phenomenon to model. But what role do ocean waves play in it?

The Role of Internal Waves

Oceanographers have made good progress in revealing the mechanisms involved in El Niño events. Theoretical studies over the past 30 years suggest a crucial role for two unusual types of ocean waves: Rossby waves and Kelvin waves. These waves were predicted to have wavelengths of hundreds or thousands of kilometers. Unlike gravity waves at the surface, these interior waves move very slowly. They take months or years to cross the full width of the Pacific. It is the long crossing time of these waves that seems to determine the interval between El Niños.

Thanks to the efforts of a number of theory groups we now have several numerical models of how Kelvin and Rossby waves participate in the El Niño cycle. Perhaps the most complete model is the "delayed oscillator" model of Stephen Zebiac and Mark Cane (Columbia University) and another model by David Battisti (University of Washington). Their models, constructed in the late 1980s, couple the atmosphere and the ocean in a complex sequence of events. The stages of their model are shown in figure 10.2.

The basic picture evolves as follows. Imagine that we are in a La Niña year. The strong southeast trade winds drive warm surface water to the western end of the equatorial Pacific, around Indonesia. This warm water accumulates in a deep pool, whose upper surface rises about half a meter above the mean sea surface. The warm water is separated from the cold deep water by a thin layer, called the thermocline, at a depth of a few hundred meters.

Moist air rises over the warm water in the west Pacific, leaving low pressure at the surface. The air flows eastward in a large convection cell, drying out as it travels, and descends near South America. This hot, dry air produces a high-pressure area near the surface which drives the strong trade wind toward Asia. So in a La Niña period, the trade winds drive the warm water westward, and the warm waters drive the atmospheric circulation eastward, which drives the trade winds, in a closed loop.

Now imagine that a change occurs in the atmospheric pressures at the east and west coasts of the tropical Pacific that causes the southeast trades to fade in strength. As the wind's push on the sea decreases, the mound of warm water near Asia begins to slide eastward under the force of gravity. The movement of the warm water is controlled by a Kelvin wave, which is a kind of

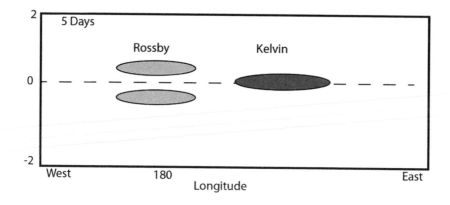

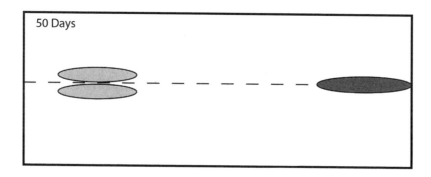

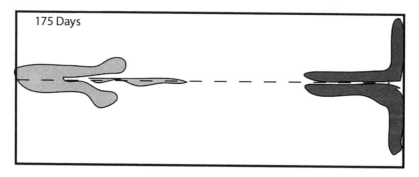

Fig. 10.2 The stages of the El Niño–La Niña cycle in the delayed oscillator model. The left and right sides of the drawing represent the Asian and tropical South American coastlines of the Pacific Ocean. (Drawn after International Research Institute for Climate and Society, "Advanced ENSO Theory: The Delayed Oscillator," http://iri.Columbia.edu/climate/ENZO /theory/evolution.html.)

long-wavelength gravity wave (fig. 10.2, top) that propagates along the thermocline. As the wave travels eastward on the equator, it depresses the thermocline, allowing the warm water to slide eastward. The Kelvin wave travels at a speed of a mere 3 m/s, so it takes about 70 days to arrive at the South American coast (the right-hand border of the figure). Its height above the mean surface of the sea is only a few centimeters, but the theory predicts an amplitude of about 50 m at the thermocline.

As the warm pool moves eastward, warm surface water north and south of the equator is sucked toward the equator. This inflow leaves a shallower warm layer in two regions on either side of the equator (fig. 10.2, top). The inflow launches a pair of Rossby waves toward the west in the opposite direction to the Kelvin wave. Rossby waves transport vortices in the water and are sensitive to the Coriolis force, as we shall see. (If you need to recall how the Coriolis force arises, take a look at the two paragraphs at the end of this chapter.) As they travel westward, at about 1 m/s (thus, three times slower than the Kelvin wave), the Rossby waves *raise* the thermocline and thin the warm layer in their paths. They also depress the surface of the sea by a few centimeters from the mean as they pass by.

So the original pool of warm water in the western Pacific is being urged toward the east by the combined action of the Kelvin wave (traveling eastward) and the Rossby waves (traveling westward). When the faster Kelvin wave reaches the eastern coast of the Pacific, the warm water it brings triggers the onset of an El Niño event (fig. 10.2, middle). Then the Kelvin wave divides. One part moves north along the coast; another part moves south. In addition, part of the Kelvin wave is reflected back as a slow Rossby wave, which propagates westward along the equator (fig. 10.2, bottom). This third Rossby wave acts to depress the thermocline and therefore deepen the surface layer of warm water. It also raises the surface of the ocean a few centimeters as it travels.

Now the atmosphere enters the picture. The warm surface water, newly arrived at the coast of South America, causes the overlying air to rise by convection. As a result the air pressure at the surface decreases, and that change further weakens the trade winds. So the development of the El Niño leads to changes that enhance the El Niño effect. This is a positive feedback mechanism that would tend to lock in the El Niño permanently. But there is more to come.

After 125 days (in the model) the Rossby wave in the west reaches the Asian coast and is reflected eastward as a Kelvin wave. But the reflection causes the new Kelvin wave to act differently: now it *pulls up* on the thermocline. Figure 10.2, bottom, shows this reflection at 175 days. When this Kelvin wave arrives at South America, at 275 days in the simulation, it thins the warm layer, allowing cold deep water to well up. The warm water is displaced by cold water. With cold water near the surface, the El Niño event ends. The original pattern of high atmospheric pressure in the Pacific's far west and low pressure in the east is slowly restored, and a La Niña phase begins again. As David Battisti put it, the El Niño effect bears the seeds of its own destruction.

This delayed oscillator model correctly predicted the sequence of events, but the times required for waves to cross the Pacific were far too short. Such defects have been addressed in recent work by adding the role of ocean currents and atmospheric waves.

Confirmation of Internal Waves by Satellite

In the early 1990s, these models were controversial. Rossby waves were familiar phenomena in the polar atmosphere (as we will see a bit later in the chapter), where they play an important role in generating the jet stream and the cyclonic weather fronts. But neither Rossby nor Kelvin waves had ever been observed in the oceans. The reason was that most of their action takes place at the boundary between the warm surface water and the colder water at depths of around 100–200 m. Moreover, their surface amplitudes were predicted to be only a few centimeters. No known technique was capable of searching for them at that time.

Then in August 1992, the French space agency, CNES (Centre national d'études spatiales), launched the Topex/Poseidon satellite from its base in equatorial French Guiana. (The name comes from "Topography Experiment" combined with the name of the god of the seas, Poseidon.) The satellite was a joint project of NASA and CNES, conceived at a seminar in 1981. NASA supplied the primary instruments, and CNES built the craft and launched it with its Ariane rocket. Walter Munk, the preeminent oceanographer whom we have already met in previous chapters, described this satellite as "the most successful ocean experiment of all time."

This spacecraft was the fourth satellite designed specifically for research in oceanography, and it was the best equipped by far. Its primary mission was to measure the hills and valleys of the ocean surface (the ocean's topography) as they vary over months and years. For this purpose the spacecraft was equipped with a novel dual-frequency radar altimeter. It was capable of measuring the height of sea level with an accuracy of 3 cm from an altitude of 1,300 km. Another instrument could measure the temperature of the ocean. The spacecraft could scan more than 90% of the ocean surfaces, repeating the same track on the sea every 10 days. Topex/Poseidon operated without a flaw and generated a mountain of data until a malfunction in 2005 ended its life.

From the measurements of sea level and temperature, scientists were able to follow changes in the major currents of the oceans, such as the Gulf Stream. The combined observations yielded estimates of the heat transported by the currents, a crucial quantity for climate research. In addition, the spacecraft was able to monitor the development of an El Niño cycle, so important for the global climate.

Early on, two keen researchers at Oregon State University, Dudley Chelton and Michael Schlax, realized that they could process the massive data set for signs of Rossby and Kelvin waves. After three years of continuous operation, the satellite had accumulated enough data for a comprehensive analysis. All the fast, random motions could be averaged out, and the predicted centimeter-height waves might be revealed. On April 12, 1996, Chelton and Schlax published their discovery of Rossby and Kelvin waves in Topex data. Their results created a sensation among oceanographers.

Topex detected the alternating positive and negative sea-level pattern and the western propagation that theory had predicted for Rossby waves. The wave periods ranged from 6 to 24 months, and wavelengths varied from 10,000 km in the tropics to 500 km at latitude 50 degrees.

In figure 10.3 we see the evolution of a Rossby wave in two global sea-level maps, dated April 13 and July 31, 1993. The height of the sea varies from point to point by less than 4 cm. In both maps, the white curved line marks the trough of a wave that is traveling westward. From the latitude variation of its speed, it was identified as a Rossby wave. The speed and direction of another wave trough, traveling eastward on the equator (marked by an X on the map), suggested that it was a Kelvin wave.

Topex saw a weak El Niño event in 1994–95. This was sufficient to confirm the basic sequence of events as described by the delayed oscillator model. But there was a problem: the observed wave speeds were twice as fast as the

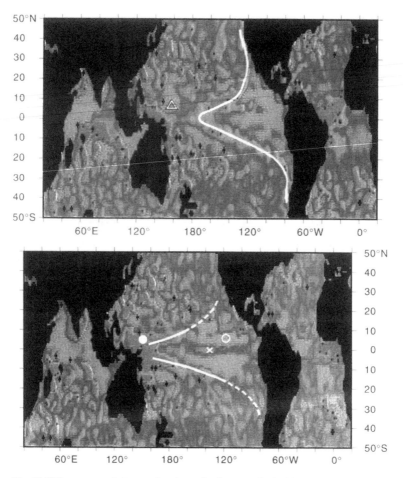

Fig. 10.3 Two maps of the sea level, made three and a half months apart by the Topex/Poseidon satellite. The curved line and light gray shading mark the trough of a westward-traveling Rossby wave. (Courtesy of D. B. Chelton and M. G. Schlax, *Science* 272 [1996]: 234, fig. 4.)

standard linear theory predicted. Since 1997, theorists have introduced a host of factors to explain the discrepancy but there is no consensus as yet.

A Comparison of Rossby and Kelvin Waves

Now that we have some idea of the importance of Rossby and Kelvin waves, let's examine them more closely. We'll begin by comparing their basic properties.

Both Kelvin and Rossby ocean waves propagate along the thermocline, just 50–200 m below the surface. Strange to say, Rossby waves travel only from east to west and at the equator; Kelvin waves travel only from west to east. But as we saw earlier, each wave can transform into the other when they are reflected back at a coastal boundary.

Both are transverse waves, in which water blobs oscillate in a direction perpendicular to the direction of propagation. In both waves, gravity acts as the restoring force in the vertical direction, as might be expected, but interestingly, the Coriolis force acts as the restoring force in the horizontal direction. It is this unfamiliar force that dictates the waves' east-west movements at the equator.

Kelvin waves oscillate primarily vertically and transmit energy like normal gravity waves. Rossby waves, on the other hand, oscillate primarily in the horizontal direction. They transport changes of rotation (or "vorticity") in the form of eddies. Both types of waves oscillate vertically by a few tens of meters on the thermocline and by a few centimeters at the sea surface.

Both have long wavelengths, up to thousands of kilometers, and extremely slow phase speeds. Kelvin waves travel at about 3 m/s and cross the Pacific in about 10 weeks. Rossby waves travel at a snail's pace, about 1 m/s, and take about 7 months to cross the Pacific.

Kelvin Waves

In 1879, more than a century before internal waves were confirmed by data from the Topex satellite, Lord Kelvin described a wave that oscillated under the joint forces of gravity, water pressure, and the Coriolis force. Kelvin waves were later named after him. There are two varieties: coastal trapped waves and equatorial trapped waves. Each of these types is divided into surface waves and internal waves. Surface waves oscillate vertically through the full depth of the water, with decreasing amplitude at greater depths, much like ordinary gravity waves. Internal Kelvin waves oscillate differently above and below the thermocline.

A coastline prevents a Kelvin wave from oscillating horizontally. At the same time the Coriolis force presses the wave against the coast so that the wave can only propagate parallel to the coast. In this sense the waves are trapped. So the blobs in a surface Kelvin wave oscillate in vertical planes that are parallel to the coast, much like an ordinary gravity wave. The wave amplitude decreases very rapidly in the offshore direction, meaning that the wave is a thin ribbon stretched along the coast. The speed of the wave is the same as for a shallow-water gravity wave and varies as the square root of the water's depth. For example, on a coastal shelf 30 m deep, a surface Kelvin wave would travel at a brisk 17 m/s, or 62 km/h.

Internal Kelvin waves are similar to surface Kelvin waves except that their motions are different above and below the thermocline. At a coast, they are trapped by the Coriolis force in the same manner as a surface wave and can only travel parallel to the coast. However, their speed is determined by the difference of water density across the thermocline, a difference of about 0.3%. As a result, the speed of an internal Kelvin wave is much smaller (about 3 m/s) than that of a Kelvin surface wave.

At a coast the Coriolis force ensures that a Kelvin wave can travel only in preferred directions. So, for example a wave must travel toward the equator along the western coast of an ocean and poleward along an eastern coast. Therefore, a Kelvin wave can travel counterclockwise around the north and south borders of an ocean in the Northern Hemisphere and clockwise in the Southern Hemisphere.

At the equator, the Coriolis force reverses direction abruptly, from a deflection to the right, to a deflection to the left. That means that both northern counterclockwise Kelvin waves and southern clockwise Kelvin waves are trapped in a narrow east-flowing band.

Rossby Waves
Planetary Spin, Coriolis, and Eddies

Rossby waves are named after Carl-Gustaf Rossby, a Swedish-American meteorologist who was one of the first to model the large-scale motions of the atmosphere, using fluid mechanics. He was a pupil of the great Vilhelm Bjerknes, the Norwegian scientist who developed the theory of weather fronts, around 1913.

In 1939 Rossby noticed that the high-speed jet streams that circle the North Pole meander southward in wide excursions. The meanders seemed to have a wavelike structure, with two to four maxima around the perimeter of the wind system. The meanders also propagated eastward along the jet, like a wave, at much slower speeds than the air within the jet stream. Rossby recognized that the meanders were wide enough and slow enough for the Coriolis force to act as the restoring force for a wave. So he analyzed the properties of the meander and gave them a physical and mathematical explanation. His work has become an essential part of modern-day forecasting of weather fronts.

Rossby was apparently unaware that his type of wave had been predicted in 1897 by Sydney Samuel Hough, a mathematician at Cambridge University. Hough had discovered the wave during his investigation of the tides on a rotating sphere. (We will come to the spiral pattern of tides again in chapter 11.)

Rossby also speculated in 1939 that his type of wave might be observed in the oceans. He would have been delighted to see the Topex/Poseidon experiments confirm that, indeed, these planetary waves are as important in the oceans as they are in the atmosphere.

The Generation and Propagation of Rossby Waves

Now let's see how Rossby waves are generated and how they propagate. We'll focus on the horizontal motions in the wave.

Imagine a column of water a few hundred kilometers in diameter somewhere in the North Pacific. It is slowly rotating in an arbitrary direction about its vertical axis. Let's call this rotation its *relative* spin (i.e., relative to the static water around it). An observer floating in space would see that the column has an additional spin. It is rotating as a whole about the earth's axis, as the earth turns from west to east. We'll call that rotation its *planetary* spin.

The size of the column's planetary spin depends on its latitude, or, more accurately, the angle between its vertical axis and the axis of the earth. If the column is near the North Pole, its vertical axis aligns exactly with the earth's, and so it absorbs the full planetary spin of the earth, a counterclockwise 360 degrees per day. If the column is near the equator, its axis is perpendicular to the earth's axis, and so it receives no planetary spin. That is, the planetary spin of a column increases with increasing latitude, in either the Northern or the Southern Hemisphere.

A most important principle, discovered by Lord Kelvin, is that the *total* spin (planetary plus relative) of a column of fluid is constant, independent of its movements in latitude. This concept is called the conservation of vorticity (or spin).

Now imagine that a sustained wind pushes our column of water north from its present latitude in the Northern Hemisphere (fig. 10.4A). As a result, the column enters a region of larger planetary spin (fig. 10.4B). Because its total spin has to remain constant according to Kelvin's principle, and because its planetary spin has increased, its relative spin must decrease by an equal amount. This decrease of spin represents a braking action, which generates a clockwise rotating eddy in the top of the column (shown as arrows in fig. 10.4B). This shallow eddy pulls water northward on its west side and pulls water southward on its east side.

The water moving north loses relative spin in the same way as before; the water moving south gains an additional amount. The result is that the rotating eddy moves to the west (dashed circle in fig. 10.4C). Notice that the water in most of the column doesn't move west; only the clockwise rotating eddy does. In effect, the change in relative spin (or vorticity) is being transmitted to the west by a traveling eddy. We can think of this eddy as part of a Rossby wave that carries information to the west.

At the same time, the Coriolis force pushes on the northernmost and southernmost segments of the clockwise eddy (fig. 10.4B). Because the Coriolis force is larger at the northern segment, the net force is toward the south (as shown), and therefore the column is driven south toward its original position (fig. 10.4C).

When the column passes *south* of its original position, another eddy of the Rossby wave is generated. Because the planetary spin *decreases* in this move south, the relative spin *increases*, and therefore the eddy rotates counterclockwise (fig. 10.4D). The net Coriolis force on the column is northward. In short, the Coriolis force acts as a restoring force, maintaining a north-south oscillation of the column after an initial push by a wind.

So we can visualize the Rossby wave as the wake of a kayak that is being paddled westward, with alternating eddies on either side where the paddle has stirred up the water.

Rossby waves can travel westward along any parallel of latitude, unlike Kelvin waves, which are trapped at the equator and along the coasts. It turns out that the speed of a Rossby wave depends on the latitude gradient of the Coriolis force, so the speed is smaller at higher latitudes. You can see this effect in the Rossby wave of figure 10.3. The trough of the wave bends backward at higher latitudes. At the same time these horizontal motions are in play, the water is oscillating vertically as well under the force of gravity, causing the thermocline to undulate up and down.

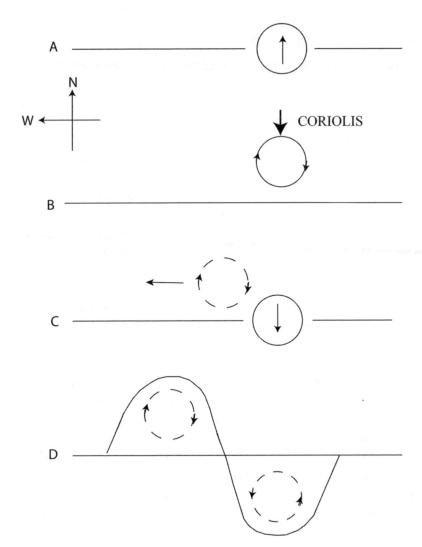

Fig. 10.4 The origin and propagation of a Rossby wave. The horizontal line represents the initial latitude of a column of water. *A*, the column is pushed farther north from original location; *B*, the column enters a region of greater planetary spin and is subjected to the Coriolis effect; *C*, the rotating eddy (*dashed circle*) moves to the west, while the column of water returns to original position; *D*, another eddy is created when the water column moves south.

The Importance of Rossby Waves

Rossby waves have now been detected in all the oceans of the world. They are recognized as the prime mechanism for transmitting changes in the ocean's circulation in response to changes in atmospheric pressures and winds. For instance, they transmit changes in the tropical oceans to higher latitudes and modulate the Pacific Ocean over periods of a decade or more. They also intensify western boundary currents, such as the Gulf Stream and the Kuroshio Current off Japan, with important consequences for the climate of Europe and North America. Recent research has shown that these slow-moving waves play an essential role in global climate change.

Rossby waves also may play an unexpected role in feeding the tiny phytoplankton of the tropical seas. These miniscule algae contain chlorophyll, which allows them to store solar energy in the form of sugar. Phytoplankton are the basic producers of food in the ocean; all other creatures depend directly or indirectly on them. The little plants require nutrients to grow and multiply, and in their vast numbers they can rapidly deplete water of nutrients. How are these nutrients replenished? One proposal is that the turbulence that accompanies currents churns the sea and brings up nutrients from the deeps. But this mechanism seems to be too weak to fit the facts.

Recent research suggests that Rossby waves may help in "plowing" the sea, turning over the water and bringing up nutrients (the "rototiller effect"). The evidence is a correlation between sea level (as measured by satellite altimetry) and the amount of green color in the water. Depressions of sea level are proxies for Rossby waves, and the green color is an indicator of chlorophyll. So the idea is that Rossby waves produce sufficient upwellings of cold, nutrient-rich water to feed the algae. However, new data upset this idea, as we shall see later.

Eddies Galore: A New Picture of the Oceans

In the past five years, more precise satellite observations of the sea level have revealed a huge, previously undetected population of eddies. Oceanographers are now challenged to interpret these observations.

We saw how Topex observations revealed, for the first time, the propagation of Rossby waves at all latitudes. Or to put it another way, that's how Chelton and Schlax interpreted the observations of ocean topography. Indeed, their interpretation was wholly consistent with the contemporary

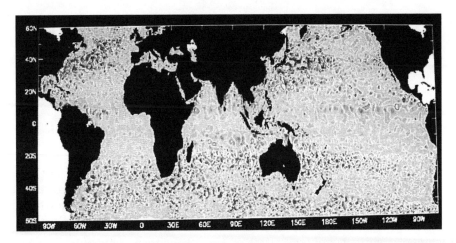

Fig. 10.5 A global map of nonlinear mesoscale eddies, obtained from observations of sea level over 16 years by two satellites. (Courtesy of D. B. Chelton et al., *Progress in Oceanography* 91 [2011]: 167, fig 1.)

theory. Their work was hailed as a breakthrough—except for the seemingly small problem that the observed propagation speeds were twice as fast as predicted.

A number of theorists immediately pitched in to explain this puzzling factor of 2 in speeds. Several ingenious ideas were proposed, and there seemed to be no serious difficulty in finding an explanation. Then in 2011, Chelton and his colleagues combined 16 years of sea level observations from two satellites, Topex/Poseidon and ERS-1. The additional data improved the spatial resolution of the images by a factor of 2, and an entirely new picture emerged (fig. 10.5).

These images showed that all the oceans, poleward of 10 degrees latitude, are sprinkled with eddies that had not been resolved before. The eddies are either depressed or raised above mean sea level by about 10 cm. They range in size between 100 and 200 km and, with the exception of one major stream, they travel nearly due west. They turn either clockwise or counterclockwise in the same hemisphere, and they have lifetimes of less than 3 weeks to more than 12. Some 30,000 eddies have measured lifetimes averaging 32 weeks.

Eddies with the longest lives could be tracked to obtain their westward speeds. These vary with latitude, from 2 cm/s at 50 degrees to 15 cm/s at 10 degrees latitude. This variation of speed with latitude is consistent with predictions for long-wavelength Rossby waves.

Chelton and co-workers conclude that these eddies are nonlinear, in the sense that their speeds of rotation are larger than their propagation speeds and that they preserve their shape as they travel. Such eddies can trap and transport seawater, an important mechanism for carrying heat throughout the oceans. Also, some eddies are green, which suggests that they, not Rossby waves, encourage the growth of algae.

Few eddies were found at latitudes below 20 degrees, where objects with much larger sizes are streaming westward at Rossby wave speeds but are not necessarily being carried by Rossby waves. However, in the higher latitudes near Antarctica, Dudley Chelton and his colleagues at Oregon State University identified a broad stream of eddies flowing westward from Australia, bouncing off the southern tip of Africa, and streaming into the *eastward* circumpolar current around Antarctica (fig. 10.5). What could this eastward streaming imply?

As you can see in the illustration, eddies seem to concentrate in dense clusters at the western boundaries of ocean basins, where major currents are found, such as the Gulf Stream; the Kuroshio, off Japan; the Agulhas, off Africa; and the Brazil, off South America. The researchers suggested that these eddies "are likely generated by the instability of background currents." These new results will provide theorists much to think about. The relationship between Rossby waves and the eddies seems to be fundamental in the dynamics of the oceans and cries out for a comprehensive explanation. We look forward to a clearer picture.

Western Coastal Currents and Rossby Waves

It is a curious fact that the current at a western boundary of an ocean basin (for example, the Kuroshio Current) is thinner and faster and carries more water than the current at the eastern boundary (such as the California Current). Rossby waves are partly responsible for this asymmetry. Let's see how this happens.

As we have seen, Rossby waves are generated by transients in the prevailing winds. These long-wavelength waves propagate westward and carry energy (and vorticity) at their group speed. When they reach the western boundary of the ocean, they are reflected. At the equator they transform upon reflection into eastward-propagating Kelvin waves, as we saw in the delayed oscillator model of the El Niño. But over most of the mid-latitudes, long-wavelength Rossby waves are reflected as *short-wavelength Rossby waves.*

So far I haven't mentioned these short wavelengths. But it should not be a surprise that Rossby waves can have a wide range of wavelengths, from hundreds to thousands of kilometers. And their group speeds and directions depend upon their wavelengths. Short wavelengths have smaller group speeds than long wavelengths and travel only eastward. That means that when a long Rossby wave is reflected at a western boundary, its reflection (a short-wavelength Rossby) carries off less energy than the incident wave brought. This phenomenon creates a steadily increasing surplus of wave energy at the boundary. To balance the energy input and outgo, the boundary current accelerates. It carries off the surplus energy in the form of kinetic energy of its waters. This process aids in the intensification of a western boundary current.

What is lacking in this simple explanation is the role of the eddies that are observed at the western boundary currents. Are these related to short-wavelength Rossby waves? Or are they related to the instability of the boundary current?

It would appear that Rossby waves are only partly responsible for the intensification of western boundary currents. Henry Stommel, a scientist at the Woods Hole Oceanographic Institute, provided an effective explanation of the phenomenon in 1948, without any reference to Rossby waves. He employed Kelvin's principle of the conservation of total vorticity (just introduced earlier in this chapter) and Sverdrup's result on the role of wind stress in accelerating a current (as we'll see in chapter 12). He demonstrated how the circulation around a gyre is generated and how the streamlines of the flow bunch together at the western boundary. This was the intensification of the boundary current. Then in 1950, Walter Munk built on Stommel's work and showed how all the gyres in the ocean fit into a global pattern of circulation. These three researchers laid the foundations of the science of dynamic oceanography.

A Note on the Coriolis Force

Imagine that a gunner and his cannon are located on the equator. He fires the cannon, intending to reach a target that lies 5 km to the north. The shell leaves the cannon at high speed aimed straight toward the North Pole. However, while the shell was held in the breech of the cannon, it also acquired the speed of the earth's rotation toward the east, which is 1,600 km/h at the equator. Now in flight, the shell retains its eastern component of velocity. As it speeds north, it passes areas at (slightly) higher latitudes whose eastern rotation speeds are less than 1,600 km/h. So when the shell finally lands, it has traveled further east than its intended target. The chagrinned gunner might think the shell has been pushed east by some unknown force.

This unknown force is called the *Coriolis effect* or *Coriolis force*, and it is purely the result of the gunner's ignorance that he stands on a rotating body like the earth. An observer floating calmly in space would see the shell going straight and would recognize that the Coriolis force is only apparent. The shell is actually following Newton's laws of motion as expected. But to an observer on earth, a body moving in the Northern Hemisphere seems to be deflected to the right, regardless of the original direction, and to the left in the Southern Hemisphere. But even though the Coriolis force is based on perception, it does have real-world impacts on the movements of objects on rotating bodies. For instance, the Coriolis force is exactly why a bathtub drains and a hurricane swirls counterclockwise in the Northern Hemisphere and clockwise in the Southern. And it also has impacts on tidal motions.

CHAPTER ELEVEN

The Tides

There is a tide in the affairs of men
Which, taken at the flood, leads on to fortune;
Omitted, all the voyage of their life
Is bound in shallows and in miseries.
On such a full sea are we now afloat,
And we must take the current when it serves,
Or lose our ventures.

That was Brutus speaking in Shakespeare's *Julius Caesar.* The bard knew full well how the image of the tide resonates in the minds of men. For people who lived by the sea, as Englishmen did, the tides were an ever-present factor in their lives. Sailors and fishermen depended on their knowledge of the tides to survive and prosper.

Twice a day a mighty wave rolls around the waist of the earth, flooding the bays and coves of every continent. Those who depend on the sea for a living have learned to anticipate the event. They know that high tide, the critical moment to launch their boats, arrives about an hour later every day. They know that the tides are linked to the phase of the moon. And they know that the height of the tide varies from one port to another.

For much of human history, that is about all that sailors and fishermen needed to know about the tides for practical purposes. But as early as 500 years ago, a few unusual men began to ask *why* the tides behave as they do.

Models of the Tides

Galileo Galilei, famous for his astronomical discoveries and his laboratory experiments, offered the first mechanical explanation for the tides in 1618. At the time, he was under tremendous pressure from the Vatican either to produce an ironclad observational proof of the Copernican system, in which all the planets orbited the sun, or to recant his support of Copernicus. Galileo recognized that if the earth moved in an orbit around the sun, as Copernicus claimed, one should be able to detect a shift between summer and winter in the positions of the nearest stars relative to the more distant stars. He had searched for such a crucial shift (called stellar parallax), but his telescopes were not sufficiently powerful to detect these tiny shifts. Therefore, he needed a different proof, and he invented a theory of the tides for the purpose.

He argued as follows. If we were to look down on the solar system from a height above the north pole of the sun, we would see the earth rotating counterclockwise on its axis, from west to east. And if Copernicus were correct, the earth would also be revolving about the sun in a counterclockwise direction. For a chosen point on earth, these two motions would add together at night, but during the day the motions would be in opposite directions and would subtract.

Therefore, Galileo claimed, the surface of the earth moves faster at night than during the day. As a result, the waters of the oceans are "left behind" at night and catch up during the day. This scheme would therefore produce a high tide at noon and a low tide at midnight at every port. Of course, it was well known that *two* high tides and *two* low tides occurred at Italian ports and that they occurred an hour later each day. But Galileo argued that the extra tides were caused by some secondary factors, such the shape of the harbors. He insisted that the occurrence of the tides proved that the earth orbits the sun and therefore that Copernicus was correct. He published his theory of the tides as part of a treatise on planetary motions entitled *Dialogue on the Great Systems of the World*.

Nowadays even schoolchildren know that Galileo's explanation was false—and he of all men should have known better. After all, he was the one to discover how the relative motions of an object combine when referred to the same fixed framework. For example, a man walking forward in a boat seems to an observer on the shore to travel faster than the boat does. But in his argument on the tides Galileo had referred the motion of the water to the earth's axis and the motion of the earth's surface to a different framework, the distant stars. The result was a muddle. Yet it is important to remember that he was a scientific pioneer doing the best he could with what he could see—and was often confronted by the immense powers of the Catholic Church to negate what he did see.

While Galileo was cooking up this stew, he was corresponding with Johannes Kepler, another giant of the Renaissance. Kepler was a German mathematician, astronomer, and astrologer—a well-respected profession in those days. Kepler had spent years measuring the elliptical motions of the planets around the sun and formulating mathematical laws to describe them. He was distracted from his work by a prolonged battle to defend his mother from a charge of witchcraft. He succeeded eventually and freed the poor woman from prison.

Unlike Galileo, Kepler recognized that the moon must play a crucial role in the tides. Somehow, Kepler claimed, the moon exerts an attractive force on the earth and pulls up a heap of water on the side of the earth it faces (fig. 11.1). As the earth rotates on its axis toward the east, the heap remains facing the nearly stationary moon. In effect, the heap is the crest of a tidal wave that travels west with respect to the solid earth.

Each port on earth would rotate under the heap, once a day. When it was precisely under the heap (like the black circle in the figure) it would experience a high tide; when it had rotated further east (like the open circle) it would be at low tide. That would account for a single daily high tide and low tide at each port, a *diurnal* tide. Moreover, because the moon moves steadily east in its orbit around the earth, the high tide would occur later every day, in agreement with observations. But just like Galileo, Kepler could not explain why most ports experience a high tide *twice* a day, a *semidiurnal tide*—or even more puzzling, why tides at some ports have two unequal maxima and two unequal minima.

Galileo dismissed Kepler's idea that the insignificant moon might raise the height of the sea from a distance. This was mere magic, "useless fiction," he scoffed. Nevertheless, Kepler had anticipated Isaac Newton in conceiving gravity as an invisible force, analogous to magnetism, that could act over great distances through empty space. That was no small achievement.

Earth   Moon

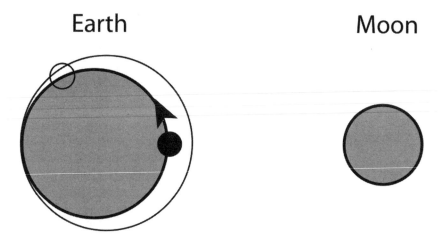

Fig. 11.1 Kepler's concept of the origin of the tides. We are looking down on the North Pole. The moon somehow raises a single heap of water in the oceans, and the solid earth rotates east through the heap. In effect the heap is the crest of a tidal wave that travels west. The port at the black dot has a high tide; the port at the white dot has a low tide. Kepler predicts a diurnal tide.

Newton's Model: Gravity and Centrifugal Forces

It was not until about 60 years later that Sir Isaac Newton entered the debate, equipped with his theory of gravitation. Newton, like so many of the new thinkers of the Renaissance, was a polymath—a physicist, astronomer, mathematician, natural philosopher, alchemist, and theologian. These studies were not considered to be antithetical to each other. Indeed, professors of universities were usually required to be ordained priests, a situation with which the very opinionated Newton was quite uncomfortable. When he eventually was elected Lucasian Professor of Mathematics at Cambridge University in 1669, he avoided being ordained until he was finally granted an exception by King Charles II.

In 1687 Newton described gravity and the three laws of motion in his famous monograph *Philosophiae Naturalis Principia Mathematica*. According to his law of gravity, every particle of mass in the universe attracts every other particle with a force that depends on the mass of each particle and that grows weaker as the distance between them (squared) increases. He could not explain the origin of gravitation, any more than Kepler could (modern physicists are still grappling with that question), but he could describe how gravity behaves.

As applied to the tides, the moon's gravitational pull would raise a heap of water on the side of the earth it faced, as Kepler had intuited. But the new feature in Newton's theory was the role of centrifugal force. As we shall see, this force produces a second heap opposite to the first one.

According to Newton's first law of motion, every object continues in its state of rest, or of uniform motion in a straight line, unless compelled to change that state by external forces. Anyone who has enjoyed a ride on a carousel has felt the centrifugal force that tends to throw the rider off the rim of the machine. That force represents your body's tendency to continue in a straight line. To stay on the carousel and travel in a circle, you must exert a counteracting force by clutching firmly to your wooden horse. A centrifugal force arises whenever a body is forced to travel in a curved path rather than its preferred linear path. The faster the speed in the path and the tighter the curve, the stronger is the centrifugal force.

Newton recognized that each planet is kept in its orbit around the sun by the balance of these two forces: the sun's gravitational pull and the opposing (outward) centrifugal force. This force arises because the planet is revolving in its curved path at a definite speed. The balance of forces ensures that the planet remains at (nearly) the same distance from the sun. (We are ignoring the existence of elliptical orbits.)

One might think that, in the same way, the moon is kept in its orbit around the earth by the balance of gravitational and centrifugal forces. But the earth is not immovable in space. The moon's gravitational pull on the earth would cause the earth and the moon to draw together unless an additional force balanced the moon's pull on earth. That additional force turns out to be a centrifugal force on the earth. How does it arise? And what does it have to do with tides?

We can see the answer in figure 11.2. As Newton taught us, the earth and the moon both revolve about a common point in space, the "center of mass," or barycenter. The line joining their centers simply turns on this center, once a lunar month. The earth is a short distance from the center of mass and revolves about it in a small circle, labeled "earth's orbit." Meanwhile, the moon is at a larger distance from the center of mass and revolves in a larger circle. In fact, the distances of the earth and moon from the center are inversely proportional to their masses. The situation is rather like a balanced seesaw, with the heavier earth closer to the pivot than the lighter moon. Or, if you like, the situation is like a heavy father swinging his light child in a

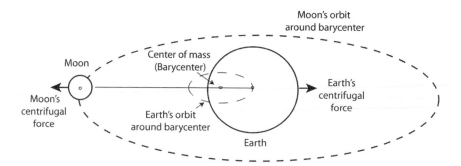

Fig. 11.2 The earth and moon revolve about a common point in space, the center of mass, once in a lunar month. The earth's motion about this point, in a small circle, creates a centrifugal force that raises a tidal bulge on the side of the earth farthest from the moon.

circle; the father has to move in a small circle to balance the pull his child exerts on his body.

So the earth revolves in its small circle once a lunar month. It is this motion on a curved track that generates a Coriolis force equal and opposite to the gravitational pull of the moon. The system of forces is then stable. To make the situation clear, I have drawn the earth's small circle much larger than it really is. In reality, the radius of the circle is 4,640 km, compared with the earth's radius of 6,371 km. So the center of mass lies inside the earth.

The Origin of Tides

To better understand tides, let's assume for a moment that the earth is completely covered with a global ocean, without continents or islands to interrupt the tidal wave. Then later, we'll examine the complications that the land introduces.

As the earth rotates about its axis, the moon's gravity pulls on the global ocean and raises the familiar bulge on the side of earth facing the moon, as depicted in figure 11.3 (here, we are looking down on the North Pole). In effect, this bulge is the crest of a tidal wave that is traveling west with respect to the ocean bottom as the earth rotates.

On the side of the earth farthest from the moon, the centrifugal force acting on earth raises a bulge in the global ocean. The moon's pull on the water on the far side is slightly weaker because that side is an additional earth's diameter (12,750 km, or 8,000 miles) farther away from the moon. This combi-

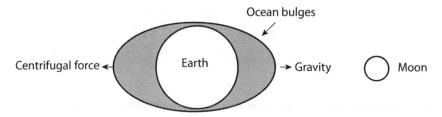

Fig. 11.3 On a world completely covered in water, the moon would raise two heaps of water on a line joining the centers of the moon and earth. (The view is looking down on the North Pole.) As the solid earth rotated through these stationary heaps, each port would experience a semidiurnal tide wave.

nation of the weaker gravitational pull and the stronger centrifugal force causes the puzzling second bulge.

As the earth rotates under the two bulges every day, each coastal area experiences two high and two low tides. The high tides occur when the coastal area lies under a bulge. The low tides occur when the coastal area lies 90 degrees from a bulge. This is the basic explanation for the semidiurnal tides.

The Variability of the Tides

Now after this dose of astronomy, we are ready to tackle the variability of the tides. For instance, why do tides shift in time? Why are some higher than others? And how are they affected by the geography of the land around them?

The moon moves 12 degrees east in its orbit during a 24-hour day. Therefore, a specific coast on the earth must rotate an additional 12 degrees, a movement that takes 50 minutes, in order to catch up with the same tidal bulge. So a tidal day—the time required to return to the same high tidal bulge—equals 24 hours and 50 minutes.

The sun also generates tidal effects. Although it is much farther than the moon, the sun is also much more massive. Hence, the sun produces a tide about 45% as high as the moon's. Its tidal bulges are permanently aligned with the sun-earth direction, as you might expect. When the moon is new or full, it lies on this sun-earth line and so, of course, do its tidal bulges. (This three-body alignment is known as syzygy.) These combined solar and lunar tidal waves, called spring tides, are higher than average. When the moon is in its first or its third quarter, 90 degrees from the sun-earth direction, the solar and lunar tides tend to cancel each other, and the tide is lower than average. This is a neap tide. So spring tides occur about every two weeks (not just in spring, a common misconception), with neap tides occurring in between, with a full-moon-to-full-moon cycle of 29½ days.

As the moon revolves about the earth, it moves about 5 degrees above and below the plane of the earth's equator. As a result, the peak of the tidal bulge varies in latitude during a lunar month, and that affects the height of high tide at different ports. In addition, neither the moon's nor the earth's orbit is a perfect circle: they are ellipses, as Kepler had calculated, and therefore they add another variation in high tides during the lunar month.

Tides in the Real World

Our real world, of course, is not an ocean planet. The continents divide the global ocean into a number of basins that influence the tides of specific coastal regions. Figure 11.4 is a 1994 map of the tidal patterns around the world that is consistent with more recent satellite observations except for some details. The map is unusual in that it focuses on the oceans rather than centering on the continents, but it shows the tides especially clearly.

The darker lines on this map are called *cotidal* lines. They mark the positions of the high tide at intervals of one hour. Everywhere along a chosen

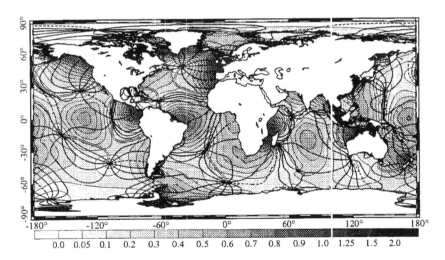

Fig. 11.4 Cotidal lines around the world. All the points on a cotidal line experience high tide at the same time. Notice the rotary patterns of cotidal lines in many places. The linear scale under the map and the dark shading in the map indicate the height of the tide. (From C. Le Provost et al., *Journal of Geophysical Research* 99, C12 [1994]: 24777, used with permission of the *Journal of Geophysical Research*.)

line the tide is high at the same time. An hour later the tide has moved on to the neighboring line. In other words, the lines indicate the positions of the crest of the tide wave, at intervals of one hour, within half the lunar day (12 hours and 25 minutes).

The rotation of the tidal waves that we see in the figure (counterclockwise in the Southern Hemisphere, clockwise in the Northern Hemisphere) is quite contrary to what we might have expected. In our previous discussion of a world completely covered with water, we saw a uniform progressive tide sweeping majestically to the west as the earth rotated to the east. In the real world most of the tidal waves rotate in an ocean basin, around a central fixed point, called the *amphidromic* point, where all the cotidal lines converge. Simple progressive waves exist only in the western Pacific, in the South Atlantic, and all the way around Antarctica. What is going on here?

Actually several things are going on at once: the gravitational pull of the moon, the rotation of the earth, the oscillation of the water in an ocean basin, and the action of the Coriolis force. We'll look at each of these in turn.

Sloshing Ocean Basins

First, many parts of the global ocean are contained in shallow basins that are defined either by depressions in the ocean bottom or by surrounding continents. The North Atlantic's basin, for example, is bounded by Labrador, Greenland, Iceland, and Western Europe. The Central Pacific is contained in a basin bounded by Asia and North and South America.

Now suppose for a moment that the moon were absent. Then the water in each basin could slosh back and forth between the bounding continents in a natural oscillation, like the water in a bathtub. A surface wave could travel to an eastern shore, be reflected to the western shore and return to the eastern shore. At one instant during such a motion, the water would slope gently downward across the basin. A high tide would occur at one boundary, with a low tide at the opposite boundary. Then the high and low tides would reverse as the tidal waves sloshed across the basin. At the center of the basin, the height of the water would hardly vary. If it were not for the friction of the ocean bottom, such an oscillation (called a standing wave, or a *seiche*) could persist indefinitely.

In this scenario, with the moon absent, each basin could have its own individual oscillation, whose period would be determined by the basin diameter and the phase speed of the tidal wave. The speed, in turn, would be determined by the fact that tidal waves are *shallow-water* waves: their wavelengths (thousands of kilometers) are much larger than the average ocean depth (about 4 km). So, as Airy taught us, the wave speed would depend only on the depth of water in the basin and would work out to about 700 km/h. (Mind you, this is the speed of energy transfer, not that of the water droplets. The currents produced by the tides rarely exceed a few kilometers per hour.) To find the period of a basin's oscillation, we could divide the circumference of the basin by the wave speed in that basin. We would find periods of these *free waves* ranging from a few hours to about 30 hours, depending on the size of the basin.

Next, let's reinstate the moon in its orbit and allow it to attract the water in a basin. Now the wave in the basin must follow the pull of the moon. It is no longer a freely running wave but a *forced wave*, whose period is equal to half the lunar day, 12 hours and 25 minutes.

So we are led to a picture of the moon driving an oscillating tidal wave in each ocean basin. But the water would still rock back and forth between opposite edges of the basin. Where do the twisting patterns come from? We have neglected the Coriolis force, which is crucial for understanding the spiral patterns in figure 11.4.

The Spiral Pattern of Tides

Let's apply the Coriolis force to a basin in the Southern Hemisphere, as pictured in the center and bottom of figure 11.4. In figure 11.5 we see how the Coriolis force affects the tides in an idealized circular basin. Here the water surface is shown in gray, the flow of water is shown by arrows, and the dashed lines represent the cotidal line of high water.

Suppose that the moon is initially located directly above the basin. As the earth rotates east, the moon drags the water to the west edge of the basin, as in figure 11.5A (at 0 hours). Therefore, a high tide forms at the western edge and a low tide at the eastern edge. Then the high water begins to flow back toward the eastern edge as the earth continues to turn and the moon's gravitational strength at the western edge diminishes.

But as the water sloshes back east, the Coriolis force deflects the flow toward the left in a *counterclockwise* swirl—that is, toward the northern edge of the basin (fig. 11.5B). This basin is so large that the Coriolis force has sufficient time (approximately 3 hours) to deflect the water by a full 90 degrees. As a result, the water piles up at the northern edge, as in figure 11.5C (at 3 hours). Now a high tide has formed at the northern edge of the basin and a

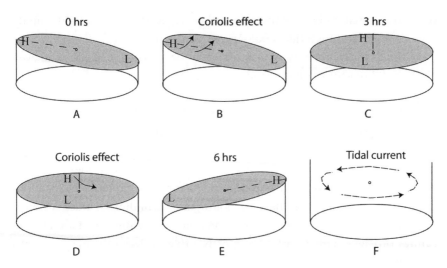

Fig. 11.5 Idealization of the spiral pattern of tides in the Southern Hemisphere. This rotary motion of high (H) and low (L) tides is produced by the actions of the moon's gravity (apparent in *A* and *E*), the earth's rotation, and the Coriolis force (moving the tides clockwise in drawings *B–D*). But the water itself flows counterclockwise (*F*) in this hemisphere. Dotted lines represent the cotidal lines.

low tide at the southern edge. The process continues: the high tide forms next at the eastern edge and the low tide in the west (at 6 hours). And so on all around the basin. Thus, the location of high tide moves *clockwise* around the basin, completing a cycle in 12 hours and 25 minutes.

In other words, the water doesn't travel directly across the basin to the opposite shore but, somewhat counterintuitively, curls toward the shore to its left (in the Southern Hemisphere). So the *water* flows in *counterclockwise* arcs while the high tide moves *clockwise* (west to north, to east, to south, to west) around the rim of the basin (fig. 11.5F). The cotidal line (shown as a dashed line) extends from the shore to the pivot point at the center of the basin. These cotidal lines are usually curved, as we can see in the rotary patterns of figure 11.4.

In the Northern Hemisphere the Coriolis force deflects water to the right. Therefore, the water flows *clockwise* and the location of high tide moves *counterclockwise* around a basin. The North Atlantic is a good example of this pattern (see fig. 11.4).

Incidentally, if the "natural" period of an ocean basin equaled half the lunar day (12 hours and 25 minutes), the tide wave would be in resonance with the gravitational pull of the moon. In other words, the moon would amplify the tide wave, and high tides could be extraordinary.

With all this complexity, we can understand why Galileo, Kepler, and even Newton did not fully comprehend the variation of the tides.

Real Tides

Tides are also affected by the geography of continental coasts, islands, and ocean bottoms. It is these realities that determine the actual heights and depths of semidiurnal tides and even whether only diurnal tides (one high tide per day) take place. Non-astronomical factors help explain why locations on the U.S. Atlantic coast have predominantly semidiurnal tides, whereas many places on the Gulf Coast have predominantly diurnal tides, and localities on the Pacific Coast have mostly mixed tides with unequal highs and lows.

In some ports, like Galveston, Texas, the full range of the tide, from high to low, is a measly 0.5 m. Near Skagway, Alaska, it is close to 3 m. In a single bay, the range of tides from one location to another can be even larger. Two ports just a few miles apart on the same coast can have very different tides, depending on the shape of their bays. It is no wonder that it took a long time to decode the tides.

The tides in certain bays around the world are truly astonishing. For example, the Bay of Fundy, which separates Nova Scotia from New Brunswick in northeast Canada, has the highest tidal range (up to 16 m) of any place on earth. Every 6 hours and 13 minutes approximately one hundred billion tons of water flows northeast up the bay. The weight of water is sufficient to actually bend down the shoreline of Nova Scotia!

The shape of the bay helps to account for these spectacular tides. The bottom of the Bay of Fundy slopes upward toward the northeast at the same time that the bay narrows in width. So the bay acts like a funnel, crowding the tide into a great heap. The bay is also special in that its natural period of oscillation is close to half a lunar day. Therefore, the tides are in resonance with the moon's tidal-raising forces, and their amplitudes are correspondingly larger. At Burncoat Head at the top of the bay, the *mean* spring tide reaches 14.8 m. And on October 4, 1869, when all the factors were favorable, the tide reached an all-time record of 16.3 m.

Not everyone is willing to grant the record of the highest recorded tide to the Bay of Fundy. Three other locations are strong competitors. Ungava Bay, which separates northeast Quebec from Baffin Island, supposedly has tides up to 17 m. King Sound in western Australia has a weaker 12-m record. And the Severn Estuary in western Great Britain has well-recorded tides up to 15 m.

In certain places the tide rolls up a river fast enough to reverse the flow of water. The tide arrives as a wall of water, a few meters high, that is called a *tidal bore*. Once again, the Bay of Fundy takes pride of place: it has spectacular tidal bores. At Truro, a few kilometers upriver of Burncoat Head, the bore height can reach 3–4 m. Although the times of the bore's arrival do not coincide either with high tide or low tide, they can be predicted in advance for the benefit of the tourists who come to watch from the shore. You can even ride the white water if you are daring enough.

Predicting Tides

Nowadays, you can look up the day's tides for any major port on the Internet, thanks to the good offices of NOAA or the U.K. National Ocean Service. Such free and accurate information is a relatively modern convenience. For many centuries, sailors and fishermen had to rely on their personal experience and the oral traditions of their community.

Then in 1687 Sir Isaac Newton published his theory of gravitation and a physical explanation for the planetary motions. Newton also described the tidal forces and the elliptical shape that they would impose on a global ocean. His "equilibrium theory" was able to explain the connection of the tides with the moon, the semidiurnal tides, and the origin of spring and neap tides. The theory also yielded sensible estimates of the tidal range.

However, Newton did not attempt to predict the tides at any specific location. Around 1740, the feisty mathematician Daniel Bernoulli picked up this task. He was born in Holland, was educated in Switzerland and Venice, and spent years working with Leonard Euler in St. Petersburg. Bernoulli made better estimates of the solar contribution to the tides and greatly enhanced the accuracy of Newton's equilibrium theory, to the point that the theory is often attributed to him rather than to Newton. He used the improved theory to compute tide tables for a number of ports, and these were the first practical, science-based predictions of tides. With such tables in hand, the captain of a ship could look up the tide at his destination and be sure he could leave it at a suitable moment.

In 1799 Pierre-Simon Laplace, a French astronomer and genius mathematician who was often referred to as "the French Newton," went one step further and derived the complete set of equations that govern the motions of a global ocean. This dynamical theory, published in his five-volume *Méchanique Céleste*, used the relatively new tools of calculus and statistical analysis to describe how the moon and the sun force the ocean to oscillate in a variety of standing waves. It explained in a mathematically satisfactory way why tides vary so much from port to port. It marked the beginning of mathematical modeling of the whole ocean based on sound physics. But unfortunately nobody was able to solve Laplace's intricate equations to obtain practical predictions.

Laplace did show, however, that the variation of the tides at many locations could be described as the superposition of three dominant frequencies, due, respectively, to the moon's orbital motion, the sun's apparent motion, and the elliptical shape of the lunar orbit. He pointed out that if one could measure the amplitudes of these three sinusoidal waves at a specific location, one could get a fair prediction of the tides.

Around 1856 William Thomson, later known as Lord Kelvin, picked up on Laplace's idea. As we learned earlier, Kelvin made fundamental contributions to hydrodynamics, to the theory of heat, and to the science of thermodynamics. Now he proposed a practical way to calculate the tides. First, he would measure the variation of the height of the water at a chosen port over several weeks. Then he could analyze the resulting record to obtain the amplitude and phase of each sinusoidal component in the time recording. In principle he could determine many more Fourier components than three. With these basic constants in hand, he could then calculate the tides for the indefinite future by superposing sine waves with the proper amplitudes and phases.

There still remained the onerous task of calculating the tides by combining Fourier components. Here, Kelvin's talents as an engineer and physicist came to the front. Kelvin was not merely an academic theoretician. In the 1860s he was employed as technical advisor to a company that was attempting to lay the first transatlantic telegraph cable. He calculated the propagation of electrical waves through a metal core and determined the limiting data rate for different sizes, compositions, and configurations of cables. In the process, he invented the mirror galvanometer to detect Morse code dots and dashes. After many philosophical battles with the chief engineer, and after the multiple technical challenges of laying 3,000 km (1,870 miles) of cable in deep ocean waters, Kelvin's engineering prevailed; and in 1866 the cable finally starting sending data between Europe and the United States at the in-

Fig. 11.6 Kelvin's machine for computing the tides at a specific location, for the entire year. (Courtesy of the Science Museum, South Kensington, London.)

credibly rapid rate of 8 words a minute. It was for this achievement that William Thompson was knighted by Queen Victoria in 1866, becoming Sir William Thompson. Later, in 1892, he was awarded the title of Lord Kelvin.

In the early 1870s Kelvin invented a mechanical analog computer that would calculate the tides, given the Fourier parameters (fig. 11.6). Kelvin had his first model built in London in 1876—a machine built of brass and steel wire that was the size of a grand piano. It could combine ten Fourier components in an accurate representation of the tide at a specific location. Its various gears and pulleys could be adjusted to incorporate the amplitudes and phases of several Fourier components, with each gear rotating at a speed that reproduced the frequency of a single Fourier component. Once the machine was set in motion, it would draw the height of the water on a moving sheet of paper and could print out all the tides at one port for a year in only 4 hours. It was a triumph of mechanical engineering.

Ten years later, William Ferrel, at the U.S. Coast and Geodetic Survey, independently conceived a similar analytical process and a similar machine to compute tides. His brass computer incorporated 19 frequencies but only predicted the times of high and low tides. A second machine that combined 37 Fourier components was built in 1912 by the U.S. Geological Survey. And a 40-component machine was built for the Liverpool Tidal Institute in 1911. Such machines were put to work to predict the tides for a full year at many ports around the world. Tables of the times of high and low water at these ports were published annually.

Predicting Tides in Wartime

If knowledge of the tides is important to seamen in peacetime, it is equally important to military forces during wartime. Here are two examples. In November 1942 the Allies invaded French Morocco with amphibious landings near Casablanca. To time their assault with high tide, the invaders needed accurate predictions of the tides for several days around the optimum date. The U.S. Coast and Geodetic Survey supplied these data, using their 30-year-old brass computer.

Then in 1944 the Allies were gearing up for the amphibious invasion of Nazi-held France. The U.S. Navy and the British Admiralty were concerned to limit the number of casualties in the assault. That meant limiting the width of beach that the attacking marines would have to cross under withering fire from the defending Germans. And that meant timing the assault with a high tide at dawn, so that the landing craft could ride up the beaches as far as possible.

But there were other conflicting constraints. The Germans had salted the Normandy coast with underwater obstacles to prevent landing craft from riding up the beaches. Demolition teams would have to land first to open up corridors through the obstacles. So the Allies decided to start the attack in an early morning soon after *low* tide, to give time for the teams to do their job. The landing craft would follow, in a rising tide.

In order to predict the tides at the five chosen beaches, the British could rely on Kelvin's original brass computer and also a newer 1906 machine at the Liverpool Tidal Institute. The machines were refurbished and readied for service, but to perform their tasks, they required the essential parameters that characterize tidal variations—namely, the amplitude and phase of a dozen different sine waves.

The British lacked this information for the five invasion beaches. They had accurate data only for the beach at Le Havre, to the east of the target beaches, and Cherbourg, to the west. Simply interpolating between the two ports would yield inaccurate results because each beach had a different coastal contour and offshore slope. Therefore, the Admiralty organized a number of daring midnight expeditions by small boats and submarines to acquire the desperately needed tidal data. The teams were only partially successful and returned with only a few tidal measurements. However, with some inspired guesswork, the gaps were filled in, and the brass computers were set to work. The proof of the pudding was that, from a tidal perspective, the landings went as planned on D-Day, June 6, 1944. The fate of the Allied men fighting on those beaches is a different and sadder story.

Modern Prediction Technology

Two technological advances that began in the 1980s have revolutionized the art of predicting tides. First, digital computers, with high speeds and huge data capacity, have replaced the mechanical analog computers of Kelvin and his successors. Secondly, radar altimetry from satellites has yielded tide heights over the whole globe with a precision of a few centimeters.

SEASAT was the first satellite dedicated to measuring the hills and valleys of the global oceans with radar altimeters. Launched in June 1978, it produced a map that was precise in defining tide height to within a few centimeters, but with relatively coarse spatial resolution. SEASAT was followed by the satellites Topex/Poseidon (operational from 1992 to 2006) and JASON 1 and 2 (launched in 2001 and 2008, respectively). They all were equipped with radar altimeters and scanning aperture radars, which steadily improved height precision (to 3 cm) and spatial resolution (to 400 km).

Satellite altimetry has a severe limitation, however. A satellite observes the height of the sea at a chosen point for only a few minutes every 10 days. So while the observations are precise, they are sparse. A hydrodynamic numerical model is therefore required to fill in the big gaps in time.

The task of integrating Topex/Poseidon observations into the existing hydrodynamic models was completed in the late 1990s, largely through the efforts of French oceanographer Christian Le Provost and colleagues. These improved models can predict the height of the tide at any point on the globe, at any chosen time. These models are now used by the forecasting agencies such as the NOAA and the U.S. Navy Hydrographic Office. What would Galileo have thought could he have seen all these incredible developments?

CHAPTER TWELVE

The Currents

So far we have been discussing the different kinds of waves that exist in and on the oceans of the world. Our story would not be complete, however, without some mention of the large, persistent currents that plow through the oceans. We know some of them by name—the Gulf Stream, for example. In this short chapter we'll touch on the global system of currents that are maintained by a corresponding system of global winds. We'll also recall why these currents are important for the earth's climate.

Trade Winds

Christopher Columbus is often credited with having discovered the so-called trade winds that blow across the northern Atlantic Ocean. In his first voyage (1492) he took advantage of the northeast wind that blows from Portugal to the Canary Islands. From there the northeast wind carried him west and south to the Bahamas in a mere 36 days. (Remember that winds are labeled by the direction from which they blow, not the direction to which they are headed.)

If Columbus had tried to return to Europe along the same track, against the wind, he would have faced months of arduous tacking and might have run out of food and water before making landfall. Instead, the canny (and lucky) captain sailed north to the latitude of the Azores (about 39 degrees), where he found a providential prevailing *westerly* wind that carried him home. In this way he discovered an antiparallel wind system across the North Atlantic basin. His voyage opened a century of exploration and trade.

He may not have recognized, however, that these trade winds drive a circular pattern of currents in the Atlantic. The south-flowing Canary Current

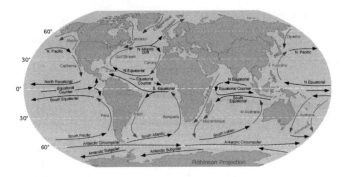

Fig. 12.1 Major ocean currents of the world. (Wikipedia.)

is one segment of the pattern; the Gulf Stream is another part. The circle is closed by the North Equatorial Current that flows westward. These currents are massive rivers in the ocean.

As you can see in figure 12.1, every ocean basin contains such a circular pattern of currents. Some currents have familiar names: the Gulf Stream, the California Current, the Peru and Brazil Currents. Others, like the North and South Equatorial Currents, or the Antarctic Circumpolar Current, are less well known to the public, at least in North America.

Gyres and Prevailing Winds

The circular patterns of currents, called *gyres*, rotate clockwise in the Northern Hemisphere and counterclockwise in the Southern Hemisphere. These major currents are permanent features of the global ocean, and they are caused by prevailing global winds. In figure 12.2 we see how the global winds change direction from one latitude band to another. So, for example, the 0- to 30-degree bands in opposite hemispheres contain the northeast and southwest trade winds. Similarly the bands between 30 and 60 degrees latitude contain the prevailing westerlies in opposite hemispheres.

If you compare the two maps, you can see how, for example, the westerly wind in the North Atlantic drives the North Atlantic Drift and the northeast trade wind drives the North Equatorial Current. The North Atlantic gyre is completed by the Gulf Stream and by the Canary Current on the western coast of Africa. Gyres in the other ocean basins are generated by associated winds in the same way.

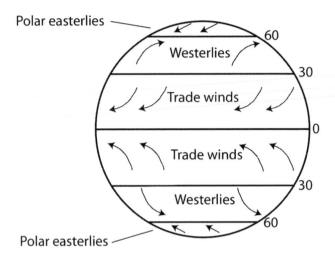

Fig. 12.2 Global wind directions vary in latitude. (Wikipedia.)

Each gyre is a slowly rotating mass of water that fills most of an ocean basin. A gyre is really a hill of water, whose center is higher than its circumference by a meter or so. Two forces combine to maintain the height difference. The Coriolis force acts on the circulating currents at the periphery of the gyre, driving water toward the center of the gyre. The water piles up at the center and tends to run downhill because of gravity. Over the long term, balance is maintained along each radius of the gyre between the Coriolis force and gravity. This steady state is called geostrophic flow.

Current Speeds and Volumes

Wind-driven currents flow at about one-hundredth of the speed of the wind, or 0.1–0.5 m/s (0.25–1.0 knots). The volume of water a current can carry is really remarkable. The Gulf Stream, for example, transports about 55 million m³/s, or 55 "sverdrups" (a measure of volume transport in the oceans). This amounts to 500 times the flow of the Amazon River. Even this giant stream is dwarfed by the Antarctic Circumpolar Current.

These major currents redistribute the heat the oceans receive from the sun, an important factor in regulating the climates of the world. The Gulf Stream carries warm water from the tropics toward Europe, which would otherwise suffer the climate of Labrador. The Canary Current returns the cooled water down south again. Similarly, the Brazil Current transports warm water south, and the Peru Current carries cold water north.

Two Oddities

The Sargasso Sea is a special sea, unique in having no coastline. It is bounded by the Gulf Stream, the North Atlantic Current, the Mid-Atlantic Ridge on the east, and the North Equatorial Current on the south. In ancient times sailors would bring home tales of ships being trapped there and being assaulted by horrific sea monsters. But the sea is actually benign and has rich populations of specialized plants and animals. Unfortunately, it has become a floating garbage dump, filled with all kinds of plastic trash that ships have spilled. (Another eyesore like this exists in the Pacific Ocean, the Great Pacific Garbage Patch.)

The Gulf Stream and the Kuroshiru Current are prime examples of a current on the western coast of an ocean basin. Such currents are stronger and faster than their mates on the eastern coasts of the basin, a phenomenon known as the western intensification of currents. In chapter 10 we saw how Rossby waves contribute to this effect.

The second oddity is the Antarctic Circumpolar Current. Sir Edmond Halley, the British astronomer of comet fame, discovered the current while surveying the Southern Ocean during the 1699–1700 expedition of the HMS *Paramore*. The Circumpolar Current is unique in having no continent to interrupt its headlong flow. It is driven ever eastward by the powerful Westerlies in the Southern Ocean. Wind speeds average between 15 and 24 knots (7.7–12.4 m/s) depending on latitude, while the current flows at less than 20 cm/s. Despite its slow speed the current transports as much as 150 million m^3/s of water, larger than any other current. Its slow speed is compensated for by a huge cross section, which extends down to 4,000 m and northward by as much as 2,000 km. The current is nearly in geostrophic balance, meaning that the Coriolis force pushes water north into a small heap, while gravity pulls the water back south, as in a gyre.

The Circumpolar Current would continue to accelerate eastward under the force of the wind if physical mechanisms did not cause it to reach a dynamic equilibrium. The nature of these mechanisms is still a matter of debate. Around 1951, Walter Munk proposed that friction with deep underwater ridges would limit the current speed at the surface. Other scientists suggest that the Coriolis force creates northward meanders of the current, so that fast water is mixed with slower water. More complex schemes have also been proposed, involving the upwelling of deep water at the southern border of the current and downwelling at the northern border.

Satellites have recently revealed a new feature of the Antarctic Circumpolar Current, a circumpolar wave. The wave travels eastward at a slower speed than the average current and completes a circuit of the globe in 8–9 years. It has two crests and two troughs. The crests are associated with pools of water 2–3°C warmer than elsewhere in the Circumpolar Current, and the troughs are 2–3°C cooler. The pools can be thousands of kilometers long.

How these waves are generated is uncertain, but they probably influence the temperature of the overlying atmosphere. There are preliminary indications that the alternation of warm and cold pools correlates well with 4- to 5-year rainfall cycles found over areas of southern Australia and New Zealand. Some scientists believe that the Antarctic Circumpolar Wave may be more important than El Niño in governing rainfall over these regions.

How the Winds Generate Currents

Now how are these major currents generated by the winds? To explain the coupling of wind and water, oceanographers have introduced the concept of *wind stress*, the horizontal force (per unit of area) that the wind exerts on the water's surface. It is a shearing force that tends to accelerate the surface of the water faster than the deeper layers.

The stress is found to increase with the square of the wind speed near the surface, as you might expect, and also depends on an observational constant. For the main purpose of describing the origin of a current, oceanographers avoid the messy details of how much the water surface is roughened by waves and bury all the details in the constant. In actuality, the so-called constant increases with wind speed, so the stress varies as the cube of the wind speed.

With this description of the driving forces and with observations of the seasonally changing wind pattern, it is possible to predict the observed pattern of global currents. In 1947, Harald Sverdrup (Scripps Institute) demonstrated mathematically how this could be done. He was drawn to the problem when he noticed that the Equatorial Counter Current in the Pacific flows *against* the prevailing winds (see fig. 12.1). The details are complex; suffice it to say that he helped to found the basic theory of ocean currents.

Of course, the major currents generate eddies on a smaller scale, so that even in the middle of a gyre, the flow patterns become quite complicated. The U.S. Ocean Prediction Center uses a sophisticated computer program, along with satellite and buoy observations, to generate daily maps of these smaller motions. They are freely available on the Web.

CHAPTER THIRTEEN

Ship Waves

If you've ever taken an exhilarating ride in a speedboat, you probably remember how the boat churned up waves as it muscled its way through the water. There were waves that rolled off diagonally to left and right from the bow, as well as those that created a foaming "rooster tail" at the stern. If you looked past the stern, you might have noticed that the bow waves form two straight lines shaped as a V that stretch far off into the distance (fig. 13.1).

Any conventional vessel that displaces water, whether a rowboat or an oil tanker, has to push the water out of its path in order to advance. That takes energy. In fact, nearly all the fuel that a boat's engine burns is used to push the water and, in the process, create waves. The waves then carry the energy away from the boat.

One of the tasks of a naval architect, therefore, is to design the hulls of ships so that they encounter the least resistance from the water and require a minimum of fuel. That means reducing the tendency of the ship to create waves. Designers of high-performance sailboats are especially concerned with optimizing the hulls. These experts use numerical simulations as well as experiments with small-scale ship models to shape the hulls. But it is still a bit of an art.

Of course, the best way to avoid making waves is to lift the boat out of the water. A speedboat or sailboat that skips, or "planes," over the water above a certain speed can do that. And a hydrofoil with underwater wings does even better. But even these craft make some waves and are subject to the same limitations.

In this chapter we'll examine the science of ship waves and the art of suppressing them. We'll see how far the experts have been able to approach their goal.

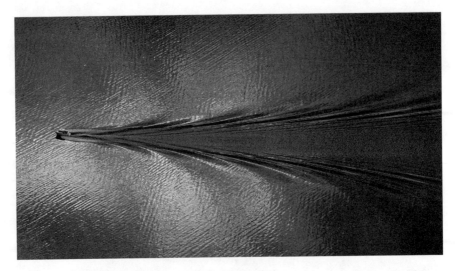

Fig. 13.1 A fast ship generates a wake of two straight lines whose angle remains the same as the ship moves ahead. (Wikipedia.)

The Classical Ship Wake

Children often draw pictures of the beautiful wave pattern that a duck makes as it paddles across a pond. Mariners are aware of the same pattern on a larger scale made by a ship on a flat sea. Figure 13.1 reminds us of the pattern. It shows the typical wake of a small, fast ship as it cruises across a bay. We see a pair of diverging waves, looking a bit like the feathers on the tail of an arrow. The turbulent wake of this ship obscures another set of so-called transverse waves, crossing the path of the ship at right angles. We see them better in figure 13.2.

If we could watch this ship for a few minutes, we'd see that the entire pattern of waves moves along with the boat. As the diverging waves spread out to the left and right, new waves are born at the bow of the boat and new transverse waves also appear at the stern. The pattern grows wider behind the boat, but its shape remains unchanged—a self-similar pattern. The pattern has the remarkable property that it always fits between two straight lines that enclose an angle of precisely 39 degrees, regardless of the speed or size of the ship. The half-angle between one of these straight lines and the path of the ship equals 19.5 degrees.

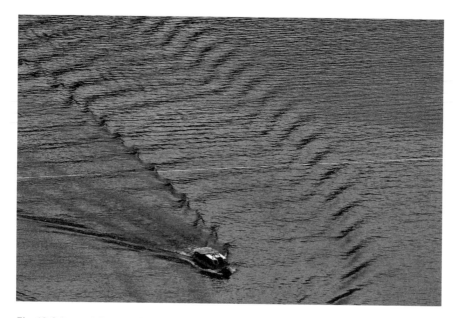

Fig. 13.2 An aerial view of a complete set of transverse and diagonal Kelvin waves in the wake of a boat. (Photo 14486777, dreamstime.com.)

Lord Kelvin and V-Shaped Ship Waves

Lord Kelvin, whom we have met several times before, was the first to explain the origin of this distinctive wave system. Kelvin was fascinated by water waves for much of his life. As a student at Cambridge, he was a member of the rowing club and was obsessed with winning trophies. His team won the annual sculling race, the Colquhoun Sculls, in 1843. Like so many scullers, he must have noticed the smooth waves his slim shell cut into the water.

Kelvin was also an avid yachtsman who loved to race his boat, *Lalla Roohk*, during the summer. In 1871 he took Hermann von Helmholtz, the famous German physician and physicist, racing at Invernay and cruising to the Hebrides. The two men, who independently discovered the Kelvin-Helmholtz instability (see chapter 3), might have debated the origin of the yacht's waves as they glided through the sea.

In addition to being a brilliant mathematician and physicist, Kelvin was also an engineer. His skill was put to a practical test when he was asked to explain why the turret ship HMS *Captain*, the glory of the British Royal Navy, sank in 1870 with almost 500 lives lost (including its designer, Cowper Coles) within six months of its launching. This wooden sailing ship was constructed with new technology: two gun turrets that could swivel around to fire in any direction. However, when the ship rolled in a gale, it quickly lurched over past 18 degrees and capsized.

The British Admiralty swallowed its pride and asked Lord Kelvin to determine exactly what had happened. He performed the calculations that showed just how unstable and top-heavy the ship had been. Even though some ship-building experts had warned against installing the heavy turrets so high up above the main deck, the political pressure was too strong, and this wooden sailing ship was built with tall masts, acres of sail—and heavy iron guns. Kelvin calculated that the ship simply had not had enough "righting moment" to roll back up upright when its deck tipped down enough to touch the ocean.

In the mid-1880s Kelvin turned his attention to ship waves. He developed a method of calculating the interference pattern that a small ship generates on deep water. He described his theory at a public lecture at the Institution of Mechanical Engineers in Edinburgh in 1887 and published a refined theory in 1906. His mathematics is daunting, but we can understand the essence of his result by using our present understanding of gravity waves.

To begin with, we should recall the pattern a supersonic aircraft generates in flight. Figure 13.3A shows how the craft excites a spherical shockwave at every point along its path. Each wave expands from its fixed center at the constant speed of sound (1,236 km/h, or 768 mph, at sea level). If the plane flies at a constant speed, all the spheres will arrive at a common three-dimensional shock front that has the shape of a cone.

As the plane flies on, the entire cone moves with it (fig. 13.3B). New shocks appear just behind the plane (shown in dashed lines), and the existing shocks grow in size. We don't see the individual shocks inside the cone because they overlap and cancel each other out. Only at the conical front are the shocks in phase and therefore visible. The angle at the tip of the cone is determined by the ratio of sound speed to aircraft speed. So the faster the plane flies, the narrower the cone.

Now let's turn to ship waves in deep water. They are similar to airplane waves in some respects and different in others. If we think of the ship as a tiny point on the surface of the sea, moving at a constant speed, it will generate circular gravity waves with a broad wavelength spectrum at every point along its path, in analogy to the supersonic aircraft. It would be difficult to

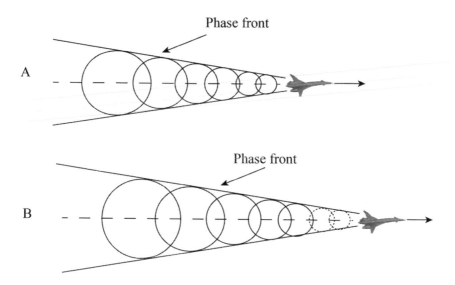

Fig. 13.3 A: A supersonic aircraft generates a spherical shock at each point in its path that expands outward to a common conical front. *B*: As the plane flies on, the pattern seen from the plane always looks the same. The faster the plane flies, the smaller the angle of the V shape.

describe how all these waves, with all their different wavelengths, interfere to produce the V-shaped wake with its unique angle. The interference pattern at any moment in time would have contributions from all the waves that had already been generated. Kelvin accomplished this difficult task using a power-ful mathematical principle that he discovered. But we can understand some of the main ideas if we consider one wavelength at a time.

Unlike shock waves, which all travel at the unique speed of sound, gravity waves in water have wavelength-dependent phase speeds; the longer the wavelength, the faster the phase speed. Also, as we saw in chapter 2, each grav-ity wave can combine (that is, interfere) with waves of nearly the same wave-length to form a chain of wave *groups*, like beads on a string (see fig. 2.4).

Groups are important in this wake problem because they, not individual waves, transport energy away from the ship. Let's agree to call the average of the wavelengths that form a group the "dominant" wavelength of that group. Then we can say that a group carries energy at a "group speed" that is half the phase speed of its dominant wave.

Now let's consider the bow wave formed by a dominant wavelength. At each point in the ship's path a circular wave group is born with this domi-

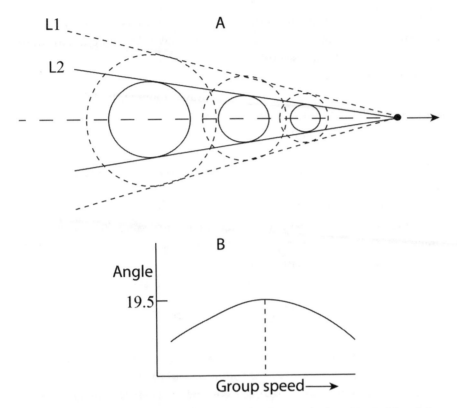

Fig. 13.4 A: A ship generates circular waves with different wavelengths at each point in its path. The waves expand at their corresponding group speeds. Therefore, each wavelength generates a front at a different angle. *B:* The half-angle of the wake varies with group speed, or, equivalently, wavelength. Wavelengths near a critical length interfere constructively to form a strong bow wave at 19.5 degrees from the ship direction.

nant wavelength and expands at its group speed. All the groups associated with this wavelength arrive at a common V-shaped front analogous to the supersonic shocks. Indeed, the ship wake of a dominant wavelength would look just like figure 13.3. But the angle of the wake would depend on the wavelength because the group speed varies with the dominant wavelength.

In Figure 13.4A we see how groups with different dominant wavelengths (L1 and L2), born at the same moment, expand at different group speeds and generate V-shaped fronts at different angles. Why, then, do we see a unique angle of 19.5 degrees?

The reason is that nearly all the V-shaped wave fronts overlap and cancel out. Only a few fronts, corresponding to a narrow range of dominant wavelengths, survive. We can see this more clearly in figure 13.4B. It shows how the half-angle of the wake varies with group speed: the curve has a maximum at a half-angle of 19.5 degrees. Only wave groups with speeds that cluster near the flat maximum of the curve will travel together and overlap to create a high bow wave. All other wave groups create fronts at half-angles other than19 degrees and interfere destructively. (This simple explanation follows one published on the Internet by Professor Erkii Thuneberg University, Finland.)

The 19-degree angle is fixed; it does not depend on the speed of the ship or its shape. But the dominant wavelength that corresponds to the 19-degree angle *does* depend on the ship speed. In fact, it increases as the square of the speed. For example, at a ship speed of 2 m/s (4.5 mph) the dominant wavelength would be 0.9 m, and at a speed of 10 m/s (22.5 mph) it would be 22.5 m.

Transverse Wake Waves: The Other Part of the Pattern

So far, I have discussed only the diverging V-shaped waves in the ship wake. As we have seen in figure 13.2, the complete pattern also includes transverse waves that cross the ship's path at right angles. Their wavelengths are directly proportional to the square of the ship's speed. So as before, the transverse waves will have wavelengths of about 0.9 m at a ship speed of 2 m/s and wavelengths of 22.5 m at 10 m/s.

Kelvin determined that the waves generated at each moment of time form a wedge-shaped or triangular object, consisting of the two diverging V-shaped waves and a transverse wave. Figure 13.5 shows three wedges formed at three moments in time. The smallest is the most recent. The next larger was formed one period before and the largest two periods before. All three wedges in the figure are expanding uniformly, preserving their self-similar shape. This is the way the pattern evolves in time, preserving its overall shape.

The diverging waves and transverse waves meet at points that are called cusps, as shown in the figure. And it is at the cusps that prominent wavelets form along the V-shaped wake. The wavelets have larger heights along the V-shaped wake because of constructive interference of divergent and transverse waves. Each wavelet slopes away from the ship's path at the fixed angle of 55 degrees.

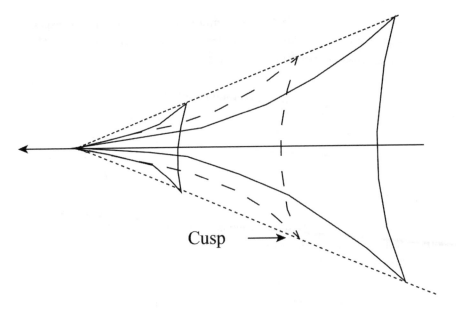

Cusp

Fig. 13.5 A ship generates transverse waves as well as diverging waves. Together, these waves form wedge-shaped patterns that expand in time as shown here. The crests (*solid lines*) alternate with the troughs (*dashed lines*).

A reality check is now in order. We've been discussing Kelvin's ideal wake, in which a pointlike source generates a wake while moving over still water. The wake of a real ship with finite dimensions has a somewhat different pattern. Moreover, the stern of a real ship also generates a wave pattern that interacts with the bow waves. We won't go into that subject, however. Instead, we turn to the practical subject of wave resistance and the struggles of ship builders to minimize this resistance.

The Battles against Wave Resistance

A ship must push its way through the water, and that takes power. Moreover, the faster the ship plows ahead, the more power it requires. To double the ship's speed, for example, the engines must deliver far more than twice the power. You have probably recognized this bald fact of life if you have listened to the engines on a speedboat. So to conserve fuel, especially at higher speeds, naval architects try to reduce the resistance a ship would encounter.

Three types of resistance are important. At low speeds the friction of the water rubbing the hull of the ship is the primary factor. At higher speeds the formation of waves in a wake becomes important. And at the highest speeds, the turbulence of the wake is the controlling factor. To distinguish "low" from "high" speed, experts now rely on the Froude number, *F*. It is a dimensionless number, equal to the ratio of ship speed to the phase speed of the gravity wave that has a wavelength equal to the ship's length.

William Froude, a naval engineer for the British Admiralty discovered that this ratio is a good indicator of the size and type of resistance a ship encounters in sailing, at least in a calm sea. The way he went about measuring resistance makes a good story.

Froude's Experiments

Froude, born in 1810, started his professional career around 1837 as a railroad surveyor and engineer. He gained notice by calculating the best curves and banking slopes for railroad tracks to prevent trains from tipping over at high speeds. Later, he was asked to look at the stability of ships, a request that caused him to become a naval architect for the rest of his life.

In the 1850s, the design of ship hulls depended solely on the experience and judgment of master shipwrights. There was very little experimental data available to guide a designer. And even the masters couldn't predict the behavior of a ship in a high wind or a roiling sea. The state of the art was highly unsatisfactory to an engineer of Froude's high standards. So Froude decided to investigate an apparently simple problem: how to minimize the rolling of a ship in heavy waves. Daniel Bernoulli, the famous Swiss mathematician and physicist, and member of the prolific Bernoulli family, had published a mathematical treatise on the subject in 1757. He concluded that the center of buoyancy (the geometric center of the underwater volume of the ship) should coincide with the center of gravity (the center of the entire mass of the ship).

Unfortunately, some of Bernoulli's assumptions led to erroneous results. Other scientists of equal ability attempted to solve the problem, but with limited success. Froude preferred to proceed with experiments first and attempt an analysis only later. He had limited funds, but he was willing to spend them to gain some insight. He began by measuring the roll of small floats in the River Dart under varying conditions of wind and wave. A typical float was a toy boat with a mast that carried a swinging pendulum. He would observe the amplitude and period of a float's pendulum and try to draw conclusions.

In 1859 he realized he needed a more controlled environment in which to conduct his experiments, so he and his son Robert built a long wave tank at his home. They fitted the tank with an oscillating paddle that allowed them to generate small waves with a chosen height and frequency. Then Froude built small ship models that they could subject to waves of different periods. Froude also varied the orientation of the model with respect to the wave fronts. With these arrangements he carried out long series of tests to determine the principles that govern the rolling of a ship. The important factors, he learned, were the positions of the centers of gravity and buoyancy, and the volume of water the ship displaced.

Froude analyzed his findings and explained them mathematically in terms of the physical forces acting on a ship. He realized he had made a significant advance in the subject and decided to publish them. But first he would expose them to the experts. So on March 1, 1861, he presented a summary of his work to the Royal Institute of Naval Architects. He received a warm response, and in the following years his work was gradually adopted by the professionals.

Encouraged by his success with ship models, Froude decided to investigate another problem facing the designer of ships: how to minimize the resistance a ship encounters in plowing through the water. The resistance was an important limitation on the maximum speed of a ship and also determined the power required to drive it. Are some hull designs more favorable, he wondered? How does the resistance vary with speed? At the time there was no mathematical theory that could provide answers, so Froude decided once again to take a purely empirical approach to the problem.

Between 1863 and 1869, in his effort to answer these questions, Froude built many models with different hull shapes and towed them in his home-built tank at constant speeds. He measured the resistance they met using an instrument of his own design. Gradually, Froude learned that the best way to determine the relationship between resistance and speed was to build two hulls with precisely the same shape but different sizes. He discovered that the resistance was the same for the two hulls if the towing speeds were proportional to the square root of their lengths. He called this relationship the "law of comparison." This result led him to devise the dimensionless Froude number; the greater this number, the greater the resistance.

In 1867 he built two ship models (called the *Swan* and the *Raven*, both now displayed in the Science Museum in London) in three sizes: 3, 6, and 12 feet. He towed them with a steam launch in the River Dart and measured the tension on the tow rope. His tests confirmed the basic concepts in his law of comparison but revealed some difficulties in the actual calculations when scaling up a model at different speeds. The problem arose because the total resistance consists of two components, friction and wave formation, which vary differently with increasing speed. But Froude found an empirical solution to this problem. And in contradiction to a statement by John Scott Russell (see chapter 4), he demonstrated that the blunt prow of the *Swan* models produced less resistance than the sharp prow of the *Raven* models.

Froude realized he could apply these results to full-sized ships. If a model had precisely the same hull shape as the ship and differed only in size, he could measure the model's resistance and scale it up to predict the ship's resistance. For example, he could build a model the same shape as that of a projected ship at a scale of, say, 1 to 20. Next, he could tow the model at different speeds and measure its resistance by the tension on the tow rope. Then he could calculate the resistance of the real ship when the ship sailed at the equivalent speeds determined by the common Froude number.

With these significant results in hand, Froude applied in 1868 to the Admiralty for funds to build a larger tow tank and to carry out a two-year research program. His request was granted in 1870, and a large tank was completed at his home in Torquay nine years later. By 1871 Froude was ready for a full-scale trial at sea. Using a scale model he was able to predict the power requirements at different speeds for the ship HMS *Greyhound*. The tests were a resounding success for his methods. The investigation into the sinking of the HMS *Captain*, chaired (as we saw earlier) by Lord Kelvin, also strongly enhanced Froude's reputation, as the investigating commission for the Admiralty concluded that the builders of this turret ship had violated Froude's 1861 criteria for rolling.

Froude went on to test hundreds of models for the Admiralty. He showed that the resistance to ship motion is proportional to the square of the ship's speed, to the square of its length, and to a coefficient, C, that rises steeply with increasing Froude number. At a Froude number of less than about 0.3, the resistance due to waves is small, and friction predominates. However, at a Froude number of about 0.35, the C coefficient increases as the sixth power of the Froude number, and wave resistance predominates. At a Froude number of 0.50 (a speed-to-length ratio of about 1.5), the ship makes large waves, and the resistance is at its greatest. The reason is that the ship's length is then about half the transverse wavelength, so that waves from bow and stern overlap constructively. This condition determines the speed limit of efficient sailing, the so-called hull speed. So to conserve fuel, most ship captains prefer to sail at less than hull speed, where the Froude number is less than 0.45.

Back in Froude's time, this hull speed seemed to limit any increases in speed of ships. But above the Froude number of 0.50, the *C* coefficient starts to decrease. That would allow new designs using very thin ends and long hulls to break through this seeming limit. Where efficiency is less of a concern, these new designs have allowed competitive kayakers, fast ferries, and catamarans to reach up to twice their hull speeds.

Froude's method of optimizing hulls with models is still used today, despite the availability of numerical modeling. The U.S. Navy, for example, operates the David Taylor Model Basin at Carderock, Maryland. It is one of the largest facilities of its type in the world, with a towing tank 850 m long and 15 m wide. The tank is used to optimize the shapes of ship hulls for maximum speed, stability, or maneuverability.

Propeller Design

William Froude and his son Robert also contributed to the design of ship propellers. In 1878 the elder Froude considered the forces acting on a segment of a propeller blade as it cuts through the water at different speeds and angles and derived formulas for the thrust a propeller delivers and the power it requires. His theory was flawed, however, because it ignored the effect of one segment on another.

Robert Froude took a different approach, which focused on the momentum of the water as it flowed past the propeller. He carried out experiments to test his ideas, with ambiguous results. Later, his theory was tested on air in a wind tunnel, and his prediction of final air speed proved to be fairly accurate. When the Froudes' ideas were combined with corrections for the interference between blade segments, a reasonably accurate theory for a simple propeller was obtained. Modern designs of complex propellers depend on a quite different approach, in which the vortices in the flow are analyzed.

Clipper Ships

It is ironic that all the while William Froude was conducting his careful experiments to understand wave resistance, shipwrights had refined a ship design that minimized resistance and maximized speed. Mind you, the design was perfected purely by trial and error, without the benefit of experiments or theory.

In the 1840s, the English middle class developed a passion for tea. To supply the product, merchants wanted a small, fast ship that could commute between India and England with a small, precious cargo. The answer shipwrights provided was the clipper ship. It was a three-masted, square-rigged ship with a narrow waist relative to its length. It carried a huge area of sail and displaced a relatively small volume of water. In other words, it packed plenty of propulsion power along with minimum drag.

The "extreme" clipper design originated in America in 1845; it was picked up and refined by the Dutch and the English. Clipper ships carried gold diggers to California in 1848 and Australia in 1851. Clippers were also used to transport opium from the Far East and mail around the world.

The clipper design reached its pinnacle with the *Cutty Sark*. She was built for the tea trade at Dumbarton, Scotland, and launched in November 1869. She measured 212 feet long, 36 feet wide and 21 feet deep. She was a slim lady indeed, with a length-to-beam ratio of 5.9 (most sailing ships of the time had ratios as small as 3). With a displacement of only 2,100 tons and a sail area of 32,000 square feet, the ship was really fast. She set a record of 73 days for the London-to-Sydney run and was capable of speeds as high as 17 knots, or 32 km/h. In the 1880s she carried wool from Australia to London. She was the last of her breed, however, as the new merchant steamships could carry more cargo at comparable speeds.

Designing a Hull

The clipper design was a triumph of practical engineering for a definite purpose; it succeeded in spite of the lack of a theory of ship resistance. It was obvious to a shipwright that if you wanted speed above all else, you reduced the displacement, slimmed down the ship, and carried lots of sail. Why these measures actually worked was not explained until John Henry Michell took up the challenge in 1898.

Michell was a professor of mathematics at the University of Melbourne in Australia and the first Australian to be elected as a fellow to the prestigious Royal Society of London. In his famous paper "The Wave Resistance of a Ship," Michell showed how to calculate the wave resistance of a "thin" ship, one whose width (beam) is very small compared to its length. He summarized his theory with a single complicated formula that involved all the parameters of the ship. His theory still yields useful results for vessels as different as racing shells, kayaks, and Navy destroyers.

In the 1980s, Ernest Tuck and his colleagues at the University of Adelaide, Australia, extended Michell's theory to ships with smaller length-to-width ratios and showed how to optimize a design. As an example of the process, suppose we want to build a sailboat for racing competitions. We want the boat to be able to exceed a certain speed, to be certain of beating its competition. We already have in mind a boat of a certain hull shape and displacement (the weight of the water that the boat will displace when it is fully loaded). To achieve our desired speed we'd like to carry as much sail as the boat can safely handle, and that depends in part on the length of the boat. (It also depends on the depth of the keel, the heavy fin under the hull.) For any chosen length we have to adjust the width of the boat to maintain our chosen displacement; longer boats will also be thinner.

With all these constraints, only the length of the boat remains adjustable. So how long should the boat be to fit our requirements and also minimize the resistance or drag on the boat?

We know that the frictional drag on the boat increases with the hull's wetted area (the area that is normally under water). If we increase the length of the boat and keep its displacement fixed, its wetted area will increase. That is, longer, thinner boats will have greater frictional resistance. But on the other hand, if we increase the length of the boat, we decrease the wave resistance. As we increase the length, holding the speed constant, the Froude number decreases, and the resistance coefficient decreases sharply.

So as we increase the length, the frictional resistance increases and the wave resistance decreases. At some length we achieve a minimum total resistance. We can calculate that optimum length with an extension of Michell's theory.

Racing Yachts

The beautiful yachts that compete in transoceanic races are the top of the line, built with titanium masts and carbon-epoxy hulls, with towering sails and lots of electronic sensors. Their performance has been fine-tuned by a process similar to the one just described. As an example consider the Open 60, a class of 18-m yachts designed to compete in the Vendee Globe races. The Vendee Globe is a nonstop, single-handed yacht race around the world that has been held every four years since 1992. The usual course begins at Les Sables-d'Olonne in France, runs down the Atlantic coast and around the Cape of Good Hope, and then rounds Antarctica in the Roaring Forties or

Fifties. The yachts then take a turn around Cape Horn and head home to France, for a total distance of more than 40,000 km.

The most exciting and dangerous leg of the course is around Antarctica, where gale-force winds and towering waves are the rule. A sailor has to be tough, resourceful, and skillful. Above all he must be able to rely on his yacht.

The winner in winter 2008–9 was Michel Desjoyeaux, who sailed his *Foncia* around the world in 84 days and 3 hours, beating his previous record win in 2000–2001 by 9 days. That corresponds to an average speed of 20 km/h, or 12 mi/h. "I had a few problems," Michel said casually at his homecoming interview. "I almost lost my bowsprit and rudders and some stanchions. I didn't sit around crying about what had happened. Once it happens, it's in the past and you move on."

Like all yachts that compete in the Vendee Globe, the *Foncia* conformed to a standard formula, the Open 60 rule, devised by the International Monohull Open Class Association. The formula constrains a combination of length, breadth, draft, and displacement, while still allowing the designer some latitude with each parameter. A typical Open 60 yacht has a length of 18.3 m, a beam of 5.7 m, a draft of 4.5 m, and a displacement of 8 metric tons. So it is far from being a "thin" ship in the sense used by John Henry Michell. Maximum speed has been compromised for resistance to capsizing, an important consideration in the Roaring Forties. Nevertheless, such a boat can be designed numerically and then tested with models, much as Froude did in the nineteenth century.

Exotic Designs

There are other ways to reduce the drag on a ship, aside from the ones I've mentioned. One way is to reduce wave generation by shaping the prow of the ship. Two extremes have been employed—a thin, sharp prow or a bulbous prow. The thin, sharp prow will cut through the water more easily, while the bulb of a bulbous prow is a blunt projection under the water that suppresses the formation of long gravity waves. Each method has its advantages at different speeds.

Another way of reducing drag is to lift some part of the ship out of the water. A speedboat that lifts its bow and "planes" at high speed is an example. Of course, in this case the wake becomes turbulent, and the increased energy losses tend to offset the gains. A catamaran or a hydrofoil achieves the same result at relatively high speeds by relying on underwater pontoons to support the main part of the craft. In particular, the low drag and lack of wake of the pontoons deliver high performance because much of the weight is below the surface of the water, while the struts that hold the catamaran deck above the water provide minimal wave resistance at the water-air boundary.

The most extreme solution to drag is the hovercraft, which floats entirely out of the water, supported only by air pressure. Here, the high power required to support the craft cancels much of the gains achieved in zero drag. It also cannot operate when ocean waves would cause loss of air pressure.

Recently, dolphins have become a topic of discussion and study among hull designers after it was estimated that they ought to have seven times as much muscle power as they actually have in order to swim as swiftly as they do, based on the hull design equations. How could this paradox be explained? Although no definitive answers have been reached, some key studies show that dolphins have minimized wave resistance through their overall sleek design and through some skin characteristics. A dolphin's skin is overall quite smooth and tough but not rigid like a boat hull. The dolphin's body and particularly its fins also contain ridges which seem to exist where turbulent flows might affect the dolphin's speed; the animal may use the ridges to control the turbulent flow around these parts of its body and thus reduce drag significantly.

The complete opposite to a hovercraft is a submarine. When submerged, a sub is subjected only to frictional drag, even at speeds that would generate substantial wakes if the sub were at the surface. So in principle a civilian submarine could make an efficient passenger vessel, capable of cruising at a steady 40 knots. It would have the added advantages of revealing the wonders of the deep to its passengers and avoiding storms at the surface. (Remember Captain Nemo's *Nautilus*?) Perhaps we'll see such craft in the future, offering a different experience in crossing an ocean.

CHAPTER FOURTEEN

Renewable Energy from Waves and Tides

Pelamis, the Sea Snake

On September 24, 2008, a group of engineers gathered on the beach at Acuçadoura, Portugal, to watch three strange devices 5 km offshore. They were the latest versions of the Pelamis wave energy converter. Over the preceding year the team had struggled to develop a reliable means of tethering the machines to the seabed 30 m below. At last the scheme was working, and the three Pelamis units were ready for a final trial. The engineers had previously proved that the machines could generate over 2 MW of electrical power. But could they safely deliver the power to the electrical grid of Portugal?

Ross Henderson, the project engineer, watched the voltage output of the machines at the substation on the beach. It fluctuated at around 690 V, close to its designed level. He crossed his fingers for good luck and closed the power switch. A meter that measures electric current sprang to life. The energy was flowing! The Pelamis system was working! It was the world's first wave energy farm. These three machines were delivering 750 kW each, for a total of 2.25 MW, sufficient to power 1,500 homes in Portugal.

The name "Pelamis" was taken from a yellow-bellied sea snake, *Pelamis platuris*, and that is what these machines resembled. Each one consisted of four cylinders connected end to end in a line 120 m long. Each cylinder was 3.5 m in diameter and weighed 260 metric tons—most of which was sand ballast to keep the cylinders partially submerged. The cylinders were joined by three universal joints that would bend in any direction. As each pair of cylinders flexed and bent in the ocean swell, a set of hydraulic rams pumped oil into a turbine that drove an electric generator. In this way, wave motion was converted to electricity. Alternating current at 690 V was conducted to the beach by an underwater cable and fed into the Portuguese power grid.

You might think that tethering a machine like this to the seabed would not hold up a team for a year, but that is what happened. In the first trials, starting in 2004 at the European Marine Energy Centre (EMEC) in the Orkney Islands, the team had developed a plug that floated 15–20 m below the surface and was tied to the seabed. The idea was to attach four tethers from the Pelamis to the floating plugs. The system would allow for quick deployment and recovery of the Pelamis without using divers. But the seabed was deeper in Portugal than in the Orkneys, and the plugs lost their buoyancy at these greater depths. Although the team solved the problem over time, the winter storms arrived before final tests could be carried out, and a whole year was lost. Nonetheless, the successful connection to the grid in September made up for the delay.

Pelamis was conceived in 1998 and developed in a decade of simulations, laboratory testing, construction, and actual ocean tests. In 2004, a Pelamis prototype in the wild seas of Orkney delivered power to the mainland power grid for the first time anywhere. Shortly after the 2008 demonstration in Portugal, the Australian investment company that had been funding the project went bankrupt; and Pelamis development, which had been going forward on a second-generation device, Pelamis P2, ground to a halt. But in 2010, the very large German investor-owned utility E.ON and Scottish Power bought Pelamis P2 generators, and development of a second-generation machine resumed.

The newest of these wave generators, Pelamis P2, consists of five connected cylinders with diameters of 4 m and total lengths of 180 m. The more sophisticated universal joints can flex in two directions, vertically and horizontally, thus capturing more of the wave's energy. As of March 2012, the machine had run for sustained periods, generating 750 kW, with a "wave-to-wire" overall efficiency of 70%. Impressive!

It is the curvature of the waves, not their height, that provides the power generated by these machines. Because the curvature varies, so too does the output of Pelamis P2. To smooth out the flow of power, the engineers have incorporated an onboard high-pressure fluid reservoir and electronics to manage the speed of the electric generator. Further smoothing can be provided with an onshore device before the power is fed to a grid.

The Pelamis must be able to survive storms, even the so-called 100-year storms. So to protect the machine, the engineers have built in a scheme that allows Pelamis to duck under the crests of high waves, the way a surfer avoids dangerous surf.

On March 19, 2012, Swedish state-owned utility Vattenfall announced a joint venture with the Pelamis company to build a wave energy farm off the Shetland coast of Scotland. The wave energy project, called Aegir Wave Power, will use 11 Pelamis P2 machines to produce 10 MW of capacity serving 8,500 households. The long-range plans call for the deployment of up to 50 MW of Pelamis power, enough to serve 66,000 homes. Such a "wave farm" would more than break even and start paying back all the research and development that has gone into this new technology.

Pelamis has moved rapidly in demonstrating a viable wave energy converter. However, its competitors are gaining ground, as we shall see in a moment. The prize is a lion's share of the infant wave energy industry. But overshadowing the future of any of these clever wave devices are the politics of climate change and fossil fuels. Let's take a moment to see whether ocean wave energy can establish a niche in the global electricity market.

The Electricity Market

The global demand for energy is now on the order of 550 quadrillion BTUs, equivalent to 160 petawatt-hours, or PWh, where a petawatt-hour is a trillion kilowatt-hours, or 10^{15} watt-hours. This enormous number represents the energy used for industrial processes, heating buildings, driving cars, and generating electricity. Overall, energy demand has been rising at an average rate of about 1.8% per year during the past 20 years.

Generation of electricity counts for about 13% of that total energy usage, at about 21 PWh per year in 2012. The rate of electric energy usage is accelerating at the rate of 2.8% per year as the populations of India and China raise their standard of living. India has already suffered major power blackouts such as the ones in July 2012 because of the lack of generation as well as a weak electrical grid. How will this increasing demand for electricity be met?

Today most of the world's electricity production derives from coal (especially in the United States and China) and oil (especially in the United States and Europe). While stocks of coal may last at least a century, the sources of easily accessible oil were, until recently, predicted to peak at mid-century and then decline. Experts gloomily forecast "the end of oil."

But this picture has yet again changed dramatically. In the world of energy, gloom-and-doom scenarios often alternate with colossal euphoria. Recent advances in drilling technology have enabled prospectors to reach pools of oil at greater and greater depths in the ocean floor. New sources of oil in the rapidly melting Arctic Ocean could conceivably stretch the world's supply of oil for a few decades. Moreover, Brazil and Africa have begun to tap their large offshore oil resources. And in the United States new technology is allowing horizontal drilling along shale layers to reach gas deposits whose recovery was previously not economical. This advance, called "hydraulic fracturing" or "fracking," promises an abundant supply of cleaner-burning natural gas for decades. Shale gas may also allow the United States to reach "energy independence" from the unstable sources of oil in the Middle East.

So on the face of it, energy production from fossil fuels now seems to have an unlimited future. Why bother with renewable technology? As most scientists now agree, the answer is that burning fossil fuels has a serious downside: the greenhouse effect. Since the Industrial Revolution, carbon dioxide from the burning of fuels has increased significantly, trapping solar energy in the atmosphere. As a result the planet is warming at an accelerating pace. Glaciers are melting, and by mid-century the Arctic Ocean may be entirely ice-free. If the Greenland ice sheet were to melt, sea level could rise by several meters. The climate of the planet is predicted to change in a variety of unpleasant directions, from more violent storms to prolonged droughts and devastating floods. So as we humans continue to burn carbon-based fuels, we are radically and unpredictably changing the earth.

Therefore, most scientists and policy makers agree that "green" renewable sources of energy must be developed to replace the burning of fossil fuels. However, there are serious disagreements on just how much money and effort should be spent, particularly as innovative technologies are improving efficiency and reducing the pollution from the fossil fuels.

Some renewable technologies are already mainstream. Hydro power from dams and rivers has been a source of power generation since the early 1900s; today, it supplies 16% of electricity worldwide. Other renewable technologies are rapidly entering the field. Wind power and solar power are two technologies that are being exploited successfully. Denmark, for example, now derives over half its energy needs from wind power. Geothermal power provides all of Iceland's power, while biomass is growing as a source (think capturing the methane from cows in a barn). Nuclear power has been considered "green" because it produces no greenhouse gases, and France derives 80% of its electricity from nuclear power plants. But after the 2011 disaster at Fuku-shima, Japan, and with the unsolved problem of the disposal of radioactive waste, nuclear power has lost many of its attractions.

The Potential of Wave Power

Three sources of renewable energy—ocean waves, tides, and currents—have barely been exploited thus far, although the ocean has often been seen as a potential energy source. As early as the start of the electrical age in 1910, the power of the waves had already been utilized by the Frenchman Bochaux-Praceique in an oscillating water column device to provide lighting to his house in Bordeaux. In the 1940s Yoshio Masuda, a former Japanese naval commander, experimented with wave energy for powering navigation lights on harbor buoys. Then when the price of oil skyrocketed in the 1970s as a consequence of OPEC's oil embargo, Stephen Salter, from the University of Edinburgh, invented Salter's duck, or nodding duck (officially the Edinburgh duck). A scale model could stop 90% of wave motion and convert most of it to electricity, at an efficiency of 81%. (We will see how this device works at the end of the chapter.) Unfortunately, the price of oil dropped equally dra-matically a short time later, so the Salter duck experiments ended.

As climate change has become more of a recognized threat, renewable en-ergy engineers have returned to the sea to try new wave energy converters: at least 22 experimental projects, including the Pelamis units, have been re-searched and tested within the last few years. Some are designed for specific stand-alone purposes like navigation lighting, but most seek to provide power to the electric grid. With the low cost of oil and natural gas, the challenge will be to make these wave generators ultimately cost effective.

The theoretical potential of ocean energy is significant. The power in a si-nusoidal wave is proportional to the square of the amplitude and to the period. Waves with long periods (7–10 seconds) and large amplitudes (say a few meters) carry about 40 to 50 kW of power for every meter along the crest. Large stretches of the coasts of North America and Europe have estimated power fluxes in this range.

One estimate is that up to 2 terawatts (2×10^{12}) of electricity could theo-retically be garnered from ocean wave, tidal, and current power. To under-stand what this number really means, it is important to understand the dif-ference between power and energy. Electric *power* measures the watts used at any point in time, while electric *energy* measures the watt-hours used over time. So a 100-W light bulb puts out 100 W of power at any moment in time and provides 100 Wh over 1 hour and 876,000 Wh over a year. Therefore, the 2 terawatts of wave power, if generated continuously, translates into about 17.5 petawatt-hours per year—almost matching the 21 PWh of electric energy usage of the entire world! Other estimates are more conservative, arriving at 2 PWh per year of practical wave energy extraction, about 10% of the world's electricity usage. Nonetheless, wave energy would be a tempting resource if the energy could be captured efficiently and economically.

The Challenges of Wave Power

Wave power has several advantages. First and foremost, it is a source of clean energy: it produces no carbon dioxide, no pollution of rivers or air, no threat of radioactivity. Secondly, wave energy is inexhaustible, although it is more accessible in some locations than others. Moreover, about 30% of Americans live within 100 miles of an ocean, and this percentage is expected to rise in the future. That means that power lines to major population centers near the coast could be shorter and cheaper than those for wind or solar facilities.

Wave power faces some serious technical challenges, however. Seawater is corrosive, a nasty environment for complex mechanical devices. The sea is also a violent environment in which to work. A ferocious storm or a hurricane could exert forces a hundred times larger than average on a device and totally wipe out an expensive installation. Even without major storms, wave devices experience continuous wear and tear from the perpetually churning waves and require constant monitoring and maintenance. Theoretical studies have to be validated by scale models, trials to test the unproven concepts, and pilot deployments to measure exactly how well the real devices perform—and last. These efforts are all very expensive.

Another difficulty is that the electrical output of a wave power device, like most renewable energy sources, is irregular in strength, frequency, and phase. Therefore, wave devices use software-driven electronic units that convert raw alternating current (AC) power to direct current (DC), which eliminates all frequency and phase issues. The DC power is then inverted back to AC but now synchronized to the grid's phase and frequency of 60 Hz (or 50 Hz in many parts of the world). However, the power output still varies, depending upon the variable wave actions. A robust power grid barely notices the variations of a small wave device (under 10 MW), but larger generators or multiple small generators have to be managed by the utilities.

Then the impact of wave devices on the ocean environment must be factored in. A few devices scattered here and there would have little effect, but more concentrated generators could affect wave interactions with beaches, the habitats of sea creatures, and the livelihood of fishermen.

So like most new technology, wave conversion involves tradeoffs. Entrepreneurs in the United States have been relatively slow to explore the technology, unlike their brethren in Europe. Let's see why.

The Realities of U.S. Wave Power

How much energy does the United States use in a year? The U.S. Energy Information Agency estimated that in 2011 we consumed a total of 30 PWh of all types of energy, out of the 160 PWh used globally. Of this 30 PWh, 15% was consumed as electricity, about 4.3 PWh. Nearly all this energy was supplied by fossil fuel and nuclear plants, with only 8% from renewables, mostly hydropower.

So what about wave power? An extensive study was carried out by the Electric Power Research Institute in 2011. The authors estimated that about 2 PWh of wave energy per year is available around the United States, including in Alaska and Hawaii, but that less than half, 1.2 PWH, is truly recoverable. Nonetheless, that is a quarter of the current U.S. electricity usage. Most of wave energy available in the United States is around Alaska (0.6 PWh), with the West Coast next (0.25 PWh). The East Coast could provide 0.16 PWh, followed by Hawaii (0.08 PWh), the Gulf of Mexico (0.06 PWh), and Puerto Rico (0.02 PWh).

The tidal power available to the United States is smaller. A study produced for the U.S. Department of Energy in June 2011 estimated that at a conversion efficiency of 15%, Alaska could supply 0.016 PWh annually. The East and West Coasts could deliver less than 0.006 PWh each.

Therefore, we cannot expect wave or tidal power to supply a large fraction of the energy demands of the nation. Only a few U.S. locations with small populations, such as the Hawaiian Islands, could conceivably meet a reasonable part of their electricity needs by using wave energy. In the case of Hawaii, whose energy use amounts to about 0.11 PWh, potential wave energy resources would be 0.080 PWh. And although Alaska has large amounts of wave power, it would not (yet) be practical to transmit it down to the lower 48 states, where the need for power is far greater.

Although wave power is not a silver bullet that can solve America's energy problems, every bit of renewable and clean energy is valuable. Indeed, President Obama's energy policy envisions using *all* available sources of clean energy, including wind and solar, in order to achieve energy independence and to restrain global warming. Then why not consider wave power?

In April 2011, the U.S. Department of Energy established three National Marine Renewable Energy Centers to explore the potential of ocean power. One was in the Northwest (Oregon and Washington States), one was in Florida, and the third was in Hawaii. The department provided one million dollars for five years, a sum which was matched by the states and their industrial partners. The Hawaiian state government, together with the University of Hawaii and several private partners, has set up a program of testing, simulation, and research. We'll touch on their test programs in a moment.

So far, wave energy technology in the United States is in its infancy. The present situation reminds me of the state of aircraft development in, say, the early 1900s, when flying machines with flapping wings were serious candidates. We have no long-term track records of the survival and maintenance costs of competing designs. We have only preliminary estimates of conversion efficiency. (For the Pelamis wave energy converter it is about 70%, but wave tank experiments on some designs run as high as 90%.) Furthermore, the sources of funding and the overall costs of building the devices and their power lines are uncertain. That means we lack realistic estimates of the ultimate cost of electricity to consumers.

So unless climate change becomes an overriding consideration, fossil fuels, with their established infrastructure, will continue to dominate the U.S. energy market for some decades to come. Nevertheless, wave energy conversion continues to attract serious attention from the government as one more tool for transitioning to a clean energy future.

The European Focus on Wave Power

In contrast to the United States, the nations of Europe are taking wave energy very seriously. In 2002, 14 European nations established the European Wave Energy Thematic Network, under the sponsorship of the European Union's Commission. This new organization coordinates the interactions among major industrial players and encourages development of alternative designs. A test site—the European Marine Energy Centre (EMEC)—was established at

Billia Croo, in the Orkney Islands, northeast of Scotland. In fact, Scotland is enjoying an economic boom as various companies arrive to test their ocean energy devices in Scotland's awesome seas.

The European Network began by estimating the wave power available at different coastal sites around the world. Along the North Atlantic coasts of Europe, the potential power ranges from a maximum of 76 kW per meter length of wave crest off the west coast of Ireland, to a minimum of 32 kW per meter length at the northern tip of Norway. If the entire Atlantic coast of Europe were covered with conversion devices, the waves could yield about 2.80 PWh of energy per year. To put this in context, the European Union consumed an estimated 3.3 PWh of electric energy in 2010. So theoretically, more than two-thirds of EU electric power could be provided by wave power.

But reality can never quite reach theoretical possibilities, and research is expensive. The cost of electricity from even the most efficient and inexpensive wave energy conversion systems is twice as much as the normal consumer's cost for electricity. Nevertheless, this result has spawned a flurry of candidate conversion schemes. In 2009 EMEC listed more than 180 companies that are developing wave energy and tidal energy devices. The industry may be young, but it is growing rapidly.

How can we account for Europe's enthusiasm for wave energy? Europeans are looking at the same facts regarding the potential of wave energy as Americans. One answer is that European policy makers are passionate about renewable energy in general. They have set the ambitious goal of meeting 20% of their energy needs from renewables by 2020. Currently the European Union's 27 member states generate an average of about 20% of their *electric* energy (but not overall energy) from renewable sources, although some are in better positions than others. The expectation is that the generation of electric energy through renewables will exceed 34% by 2020. To this end they have instituted an EU-wide cap-and-trade system to reduce carbon dioxide emissions. Wave power will be a part of the electric energy renewable goal.

Wave Energy Conversion Devices

Of the several types of wave conversion devices being proposed, there are a few basic types and one or two really unusual ones. The more common ones are these:

- The Pelamis converter is an example of an *attenuator*. It floats at the surface, pointing head-on into the waves, and extracts energy from the wave movement by flexing at the joints between its cylinders. An attenuator has a small cross section facing the wave crests, so it experiences relatively small forces. That's an advantage in a rough sea.

- A *point absorber* looks like a large buoy floating on the surface and tethered to a fixed point on the bottom of the ocean. It covers a relatively small area but bobs vigorously up and down in the waves, extracting the energy of the relative movement between the fixed ocean floor and the bobbing buoy. A point absorber is quite rugged and is capable of surviving severe storms. One prototype is being tested in Kaneohe Bay, Hawaii, by the U.S. Navy, as we shall see.

- A *surge converter* uses a large vertical paddle on the seabed near the shore that swings back and forth. As waves pass by, the paddle's motion is converted to electricity. The Oyster, discussed below, is such a device.

- The *oscillating water column* is a vertical tube in near-shore water. As the surf rolls in and out, the water first pushes the air forward to spin one turbine, then pulls the air back through a second contra-rotating turbine: together, these turbines run a generator. The device—the first type of converter to be invented—is extremely simple and rugged. One example of the principle, the 500-kW Limpet, has been generating power since 2001 at the Scottish island of Islay.

- The *terminator* uses a large enclosure. Incoming waves are funneled up a ramp into the enclosure. In effect, the enclosure becomes a reservoir slightly above sea level. The captured water then flows back to the sea through an ordinary hydroelectric turbine. The Wave Dragon in Denmark is one example.

- A *pressure differential* uses the changes in water pressure that passing waves produce near the seabed. A vertical cylinder is located below the waves. The fluid at the bottom of the cylinder maintains a constant pressure, while the fluid at the top experiences continuously changing pressure. The pressure variations drive a generator.

In addition to these, there are devices that do not fit into any category. For example, in 2008 the Stanford Research Institute, in Stanford, California, demonstrated a strange polymer that generates electricity when it is stretched. Its developers call it artificial muscle. Biological muscles, as you know, contract when an electrical signal is applied. These Stanford researchers want to reverse the process to produce electricity. Their aim is to build a device with no moving parts. The concept has been tested on a navigation buoy off the California coast; the output was small but sufficient to power the lights on the buoy. In principle, such devices could be scaled up to supply power to a grid. We shall have to see where this idea goes.

Just to give you an idea of America's effort in this field of wave energy conversion, of the 180 companies listed by EMEC as developing wave and tidal energy devices, only 27 are U.S. companies. The majority, 14, are developing point attenuator devices, similar to a buoy.

The Anaconda

Let's turn now to some innovative wave energy conversion devices that are in various stages of development in Europe. We begin with the Anaconda. It's a wave attenuator similar in shape to its namesake, the large South American snake. The prototype was unveiled in 2010 by the British firm Checkmate Sea Energy.

The Anaconda is a water-filled tube 9 m long and about 1 m in diameter, made of a rubber composite. The tube is tethered to the seabed, facing head-on into the waves and swimming just under the surface. As a wave passes by, its pressure creates a bulge in the rubber tube. The bulge travels down the tube, toward a turbine and generator in the tail. Rod Raines, a co-inventor of the Anaconda, compared the action to a pulse of blood in an artery. When the pulse reaches the tail, the generator produces electricity that is drawn off in a cable to shore.

The great advantages of the Anaconda are low manufacturing costs, mechanical simplicity, and the durability of its rubber tubing in the harsh conditions of the sea. Professor Raines envisions a scaled-up version, 200 m long, that could generate 1 MW. Power from these wave converters cost about 25 cents per kilowatt-hour in 2009, but Raines thinks the price can eventually be lowered to 9 cents as the technology for the Anaconda is refined and mass-produced. This price would compare favorably with other energy sources; for example, power from new natural gas plants costs about 7 cents per kilowatt-hour; from coal or wind plants, about 10 cents per kilowatt-hour; and from new photovoltaic solar systems, 15 cents per kilowatt-hour. We shall see if energy from wave power can be produced at similar costs.

The Oyster

The Oyster, built by the Scottish firm Aquamarine Power, is basically a wave-powered pump that delivers water under high pressure to an onshore turbine and generator. The heart of the Oyster is a large flap that is hinged at the seabed, 10–15 m down, about half a kilometer from the shore. As the waves pass by, the flap, which is mostly underwater, swings to and fro and actuates two hydraulic pistons that push water down a pipeline to the shore. One advantage of the Oyster is that it operates close to the shore, avoiding the worst of the battering by storms farther out to sea. Moreover, the most expensive machinery is located on shore, where it can be serviced easily.

Beginning in 2009, the prototype Oyster was tested successfully for 6,000 hours at the company's Orkney test site. One unit of a second-generation device, Oyster 800, was deployed in September 2011. It has an underwater flap 26 m in width and generates 800 kW. In November 2011 it was successfully connected to the Scottish grid. Venture capitalists and the Scottish government were sufficiently impressed to invest an additional £7 million for further development. As of March 2012, Aquamarine had obtained more than £70 million in grants, loans, and equity.

When complete, Oyster 800 will consist of three units, each capable of generating 800 kW. In February 2012, EMEC granted permission for Aquamarine to install the remaining pair of units at the Orkney test site. A common pipeline will feed high-pressure water to the generator onshore. Aquamarine Power has not yet published any performance characteristics, such as efficiency of conversion, for the Oyster, but it is obvious that progress is being made.

The Limpet

While many new devices are still in the early stages of development, the Limpet, developed by another Scottish firm, Voith Hydro, has been delivering half a megawatt of power to the grid on the island of Islay since 2000. It is an oscillating water column type that is mounted at the shore. The device consists of a sloping air chamber that has an opening underwater. As a wave arrives, the water pushes the air column past two counter-rotating turbines. One turbine spins as the column rises; the other, when the column falls.

The Limpet was designed to operate in relatively weak swells that deliver 15–25 kW per meter of crest width. It has the advantages of relatively simple construction and maintenance. So far it has not been copied elsewhere, although it seems to be well suited for conditions on the west coast of the United States.

The Penguin

The Finnish firm Wello Oy began developing the Penguin in 2010. The 1,600–metric ton prototype looks like a bulky barge about 30 m long. It floats at the surface, tethered to the seabed by three light cables. Inside its sealed enclosure, a horizontal eccentric flywheel is mounted on a vertical axis. As the Penguin rocks in long waves, the flywheel turns in one direction and operates an onboard generator. Power is drawn off to shore through an undersea cable. A unit rated at 0.5 MW was tested at sea during 2010–11 and was towed to the EMEC's Orkney site for extended tests in March 2012. We'll have to wait for more details.

The PowerBuoy

Ocean Power Technologies, an American company, has been developing the PowerBuoy under contract to the U.S. Navy. A 40-kW machine, which looks like a large yellow buoy, has been tested in Kaneohe Bay, Hawaii, since 2004. Beneath the "buoy" that floats atop the water, this device has a vertical cylindrical hull that extends deep under the surface. This hull barely rises and falls in a swell. In contrast, a sliding piston at the center of the hull is free to rise and fall with the waves; the difference of motions is used to actuate a generator.

A second-generation PowerBuoy, intended to be used far offshore, is rated at 150 kW. A prototype, tested for six months in 2011, 60 km off the coast of Scotland, produced 40 kW in 2-foot waves. Presumably this unit would be used locally to power a radar or navigation light, just like the little solar cells that power roadside emergency stations on many highways.

Tidal Power

What about extracting power from the tides? Tides are very predictable, unlike wind and solar renewable energy sources, and have been used for centuries to turn mill wheels. But the main problem in the past has been to convert the slow tidal motion (a wave with a 12-hour period) to electricity, which is the only way that tidal energy could be transported to where it is needed.

Clearly, simple hydro turbines can be used wherever dramatic tidal flows exist. When the tide rises or ebbs in an estuary, for example, the streaming water can be funneled through turbines to produce electrical power.

The outstanding example of such a generator is on the Rance River in Brittany, France. It is equipped with 24 huge hydro turbines with a maximum output of 240 MW of electricity. However, since the tidal water essentially stops moving during high and low tides, this generator can average only about 96 MW. Since November 1966 the facility has been generating an annual average of 0.006 PWh, a mere 0.9% of France's energy consumption of close to 0.7 PWh.

South Korea inaugurated the world's largest tidal generator at Sihwa Lake in August 2011. It has an installed capacity of 254 MW. In the United States, a 1-MW tidal station is now operating in the East River in New York City. And a novel "egg-beater" design, with a rating of 180 kW, was installed at Eastport, Maine, on Cobscook Bay in August 2012.

But there are very few estuaries that experience such dramatic flows. Would it be worth developing entire new technologies just for a small number of places around the world? Despite its limitations, many nations do see promise in tidal power. Some, for instance, have experimented with just attaching small turbines on the legs of existing bridges for very inexpensive power. Others, like Scotland, which is very rich in tidal energy, are experimenting with many different types of tidal generators. The Scottish government, with its goal of 80% renewable energy by 2020, is fully behind these experiments.

As of December 2011, at least seven full-scale tidal devices were being tested at EMEC's Falls of Warness site, off the isle of Eday, in the Orkneys. These tidal devices are being developed by countries like Ireland and Germany as well as Scotland. The site was chosen for its high-speed tidal currents, which reach 4 m/s (7.8 knots) at spring tides.

Of the devices that EMEC is testing, the most conventional is the undersea analog of a wind turbine, typically with three large blades that rotate on a horizontal shaft. These tidal turbines sit on the seabed below the turbulence of the surface waves and utilize the energy from the regular tidal streams. Since they are invisible, there is no visual pollution. Another version looks more like a weather vane, with three vertical blades mounted on a vertical shaft. Eventually, Scotland expects to deploy several types of devices in the Pentland Firth in the far north, with enough power (160 MW) to serve a medium-sized city.

A more unusual device being tested by EMEC has a hydrofoil design, in which a horizontal blade is mounted on a rocking arm. As the tidal stream flows by and lifts the blade, the arm oscillates up and down, activating a pump and a generator. This tide power generator has appropriately been named the Stingray, if you can imagine the undulating mantle of that creature.

Tidal power experts have also recognized that in addition to tidal streams where the ebbing and flowing tidal water turns the turbines, the difference in height of the water during high and low tides can also be exploited. If a dam is built across an estuary, the incoming tidal water during high tide can be captured behind the dam and then released through turbines when the tide turns. One theoretical idea being evaluated is the use of partial dams that hinder the tidal flows just enough to capture the energy from the difference in height.

Modern improvements in turbine design and construction are expected to expand the number of locations where tidal generators could be placed. In addition to the obvious locations around Scotland, other prime candidates are New Zealand, Australia, Canada, and even the Golden Gate Bridge in San Francisco.

The Tale of the Nodding Duck

Whether wave energy conversion will find a niche in the global energy market depends not only on a winning technology and low costs, but also upon politics. As an example, recall the story of Salter's nodding duck.

In 1973 the Organization of the Petroleum Exporting Countries (OPEC) declared an embargo on oil shipments, and the price of oil skyrocketed. The oil crisis caused many nations to seek energy independence from the unstable producers in the Middle East. That in turn encouraged a spurt in development of wind and solar power and, to a lesser extent, wave power. In that environment of unstable oil supplies, Salter's nodding duck was born.

Stephen Hugh Salter was a professor of engineering design at the University of Edinburgh. In 1974 he invented a wave energy conversion device, the nodding duck. It was a long flat vane with a cross-section in the shape of a teardrop, giving it a sharp leading edge and a round backside. In operation, the long dimension of the vane would be mounted parallel to the incoming wave crests, with its thin edge (the duck's beak) pointed seaward. As a wave rolled by, the duck would nod about an axle in its round back side.

Inside the duck, four gyroscopes would transmit the nodding motion to a pump that fed oil under pressure to a generator. Part of the electricity would be used to spin up the gyroscopes and therefore to store some of the energy. The stored energy could be released when the waves died down, so that the output of the duck could be fairly uniform.

Salter tested small-scale models of the duck in his wave tank in Edinburgh. He learned that his model was capable of extracting an astounding 80%– 90% of the incident wave energy. The machine essentially flattened the waves that rolled onto it. On the basis of these tests, Salter was able to patent his invention in 1975. Then Salter applied for a grant from the U.K. Wave Energy Programme to build a full-scale version and test it at sea.

In an interview in *The Engineer*, on April 10, 2007, Salter said that initially his device was received with enthusiasm: "Cost predictions were impossibly high but the people in charge of the programme were very optimistic. Their enthusiasm diminished as costs came down." Then on March 19, 1982, the government abruptly shut down the Wave Energy Programme after a critical, closed-door meeting of a selection board. Salter thinks the board was loaded with representatives of the nuclear power industry and that they voted to shut out the competition from wave power. "When the programme manager predicted that with development, the [electricity] cost would be 3.3 p[ence]/ kWh, which was getting close to being economically viable even then, they excluded him from the next important meeting of the key committee. . . . They basically killed the project because it was going to threaten the expansion of the nuclear industry." There are conflicting reports about the cost of electricity that could be produced by the duck. Some say that an error of a factor of 10 was discovered in the estimates. Some say that is implausible.

The moral of the story is that technical excellence is insufficient to assure success in a competitive market. Political skill is also needed to bring a device beyond the purely research stage. It's hard to guess whether Salter's duck would ever have been a success. In any event, enthusiasm for wave energy faded soon after the price of oil dropped to normal levels. Will that happen again this time around?

CHAPTER FIFTEEN

The Future

In the preceding chapters we have seen wave research develop to the point that reliable daily forecasts of wave heights and periods over the whole globe are readily available. Much progress has been made in understanding wave and current dynamics in the nearshore environment. Rogue waves are still not predictable, but we have come a long way toward understanding where and how they are generated. And although submarine earthquakes cannot be predicted, the propagation of the tsunamis they sometimes produce can be forecast with considerable accuracy, allowing threatened populations to be warned.

What are the likely directions of ocean wave research in the coming decades? The U.S. National Academy of Sciences organized a workshop in 2009 to ask a similar question about oceanography as a whole. At the request of the Office of Naval Research and the Naval Research Laboratory, the NAS Ocean Studies Board convened a meeting of several dozen leading scientists to try to predict which problems in oceanography would be tackled and which major themes would be addressed between 2009 and 2015. In a separate NAS workshop held in 2009 (entitled "Oceanography in 2025"), many of the same participants were asked to sketch the status of oceanography in 2025.

As you might expect, climate change will continue as a major theme in oceanography. Now and in the future, researchers will monitor and study any changes in the primary ocean currents that transport heat around the globe. They will study the interaction of the atmosphere and the sea in such large-scale cycles as the El Niño effect. They will try to measure the fluxes of gases (such as carbon dioxide), mass, energy, and momentum across the air-sea boundary layers.

In the future even more emphasis will be placed on the *variability* of winds, waves, and currents, in both space and time dimensions. The turbulent mixing of surface layers, essential for the storage of carbon dioxide and the distribution of nutrients, will come into focus. That will require new observations at the *mesoscale* of 100 km or less and timescales of a few hours. Satellites, instrumented buoys, aircraft, autonomous undersea vehicles, and unmanned drones will be improved and deployed. Ultrafast computers will be employed to extend models under the sea and to investigate turbulence at the microscale level of centimeters.

What part of this grand enterprise will wave research play?

Internal waves will become a major focus in the next decade. They were detected decades ago, first by astronauts, and then by the satellite Topex/Poseidon and with synthetic aperture radar. But only in the past decade have these waves been tracked and studied quantitatively. The transients in ocean circulation, initiated by wind pulses and modulated by internal waves, will be examined in detail. The relationship of ocean waves to mesoscale eddies has just begun to be studied. In the future, measurements of the transport of mass, heat, and vorticity, and, most importantly, the vertical mixing produced by internal waves, will receive additional attention. Peter P. Sullivan (University Corporation for Atmospheric Research) foresees that models of mixing, including realistic representations of turbulence, will become feasible with computers capable of a thousand trillion calculations per second.

Yeuming Liu and Dick K. P. Yue (Massachusetts Institute of Technology) envision a new generation of wave prediction tools by 2025. Large-scale simulations of nonlinear wave fields that incorporate satellite measurements of wave phase will be feasible by then. In addition, Liu intends to extend her present simulations to predict the short-term evolution of wave fields.

Researchers will also investigate subsurface tidal waves, which are generated near a coast or between islands. Such waves actually break underwater, producing turbulence and strong mixing.

Another major topic that will engage wave researchers is nearshore dynamics, the nonlinear interaction of surface waves and the local topography of the ocean bottom at depths less than 10 m. How do beaches grow and decay? How can erosion be prevented? How are currents modulated? These are questions of practical importance. Earlier in this book, we learned about the Nearshore Canyon Experiment of 2003. We can expect more of such intensive campaigns to explore currents, waves, and sediment transport on a complex coast. Rob Holman (Oregon State University) projected that forecasts of the short-term evolution of a specific nearshore bottom, similar to weather forecasts, will become feasible by 2025.

Finally, microscale (centimeter-length) processes at the air-sea surface associated with wave breaking will see closer examination. These processes include the formation of air bubbles, the release of salt nuclei, heat transfer, and the absorption of carbon dioxide. At the 2009 Ocean Studies Board workshop, Ken Melville, of Scripps Institution, pointed out that we still do not have a good understanding of surface wave processes. However, we do know that the "sources of mixing and turbulence on both sides of the surface are dependent on the wave field. Increasingly it is found that fluxes of mass, momentum and energy depend on the wave field, effects that formerly were hidden in the large scatter of the data." With the expected improvements in the quality of measurements, we can look forward with optimism to further understanding.

The European Centre for Medium-Range Weather Forecasts held a workshop in June 2012 to solicit ideas on improving the quality of ocean wave forecasting. Some of the topics discussed included better models of wave dissipation, wave-current interactions, and the generation of waves under hurricane-force winds. We can expect renewed efforts in these fields over the next decade.

These are some of the topics that researchers foresee as ripe for study. But the overall trend toward variability and high spatial resolution may bypass issues still not settled. Walter Munk, a distinguished participant in the 2009 Ocean Studies Board workshop, cautioned that the physics of wind stress on the sea—how the wind excites waves and currents of all types—is still not thoroughly understood.

The next decade should be an exciting one. Not only can we look forward to progress on our known problems, but there is also the prospect of new discoveries and new problems. Stay tuned!

Glossary

Amphidromic point	The point of rest in the tidal oscillations of an ocean basin.
Amplitude	The distance between the crest of a wave and the equilibrium level of the sea; for a sine wave, half the distance between crest and trough.
Barycenter	The point at which the total mass of a group of bodies can be thought to concentrate when acted upon by external forces.
Berm	A flat terrace on a beach, caused by wave action.
Blob of water	A volume of water about the size of your thumb.
Coriolis effect or force	An apparent force that seems to act on a body when observed from a rotating frame of reference.
Cotidal line	The line on which high tide occurs within an ocean basin at a chosen moment.
Cross sea	The disturbed state of the sea when a swell crosses wind-driven waves at an angle.
Diffraction	A spreading of a wave front as it passes through a narrow aperture, such as a channel between islands.
Dispersion	A variation of wave speed with wavelength. Long gravity waves travel faster than shorter gravity waves.
Energy wave spectrum	The distribution of energy among waves of different wavelengths (or periods) in a sample.
Fetch	The distance of open water over which the wind has blown.
Forced wave	A wave that moves under the influence of an external force—for example, a wind-driven water wave.
Free wave	A wave that propagates at a characteristic speed when no external forces act upon it.
Fully developed sea	The state of the sea after the wind blows steadily for a long time over a large area. The waves eventually reach a point of equilibrium with the wind.

Gradient	The change of a quantity (such as velocity or temperature) with a change in another quantity (such as horizontal or vertical distance).
Gravity wave	An ocean wave in which gravity is the dominant restoring force.
Group of waves	The envelope of the crests of several similar wave trains that overlap and interfere. A group is a pattern that travels at a slower speed than the phase speeds of its constituent waves.
Gyre	A rotating body of water that fills a whole ocean basin, turning clockwise in the Northern Hemisphere and counterclockwise in the Southern Hemisphere.
Hindcast	An attempt to reproduce the properties of the sea in a past storm, using a numerical model with wind observations as input.
Interference	The interaction of two or more wave trains that results in a pattern of enhanced and depressed wave heights.
Kilometers to miles conversion	1 kilometer = 5/8 mile, or about 0.6 mile
Knot	1.15 mph
Meters to feet conversion	1 meter = 3.28 feet = 39.37 inches
Period	The time interval between the arrival of consecutive crests at a stationary point.
Phase	The present state of a system that is oscillating in a cycle. A full moon, for example, is one of the phases of the moon.
Phase speed	The rate (in, say, kilometers per hour) that a representative point on a wave profile (such as the crest or trough) travels. The speed is determined by the wavelength or period of the wave.
Refraction	The turning of a wave caused by a variation of wave speed along the wave front. Speed variations are commonly caused by variations in the depth of the ocean bottom near a shore.
Resonance	A state in which two oscillations have the same frequency. When a driving force acts on an oscillator at its resonant frequency, the amplitude of the oscillation increases.
Ripple or capillary wave	A wave whose dynamics are dominated by the effects of surface tension.
Rogue wave	A wave higher than expected in the present state of the sea. Specifically, a wave height at least twice the significant height.

Rossby wave	A complex wave that transports changes of rotation or vorticity across an atmosphere or ocean. The Coriolis force acts as a restoring force for the wave's oscillations.
Self-similarity	The condition in which the whole is exactly similar to a part of itself: the whole has the same shape as one or more of the parts.
Shoaling wave	A wave that changes its speed and shape as a result of encountering water depths that are a small fraction of its wavelength.
Significant wave height	The mean wave height (trough to crest) of the highest third of the waves.
Slope or steepness	The dimensionless ratio of wave height to wavelength.
Slosh zone	The region on a beach where the water from a breaker runs up the slope of the beach.
Soliton	A single isolated group of waves that propagates without a change of shape; a wave "packet."
Spectrum of waves	A graph that shows how the wave energy in a sea is distributed among waves of different wavelengths or directions.
Stokes drift velocity	The speed with which a blob of water advances in the direction of propagation of the Stokes wave.
Stokes wave	A water wave of finite height that preserves its nonsinusoidal shape as it propagates despite the destructive effects of dispersion.
Storm surge	The mountain of water that storm winds push toward the shore.
Surface gravity wave	A wave formed at the interface between two media of different densities caused by the restoring force of gravity or buoyancy.
Swash zone	The area of beach covered by successive breaking waves.
Swell	A series of surface gravity waves not generated by the local wind but dispersed from a distant generation area. They often have a narrow range of long wavelengths.
Synthetic aperture radar (SAR)	A form of radar that uses the relative motion between an antenna and its target region to provide distinctive spatial signatures that are exploited to obtain finer spatial resolution than is possible with conventional beam-scanning means.
Tsunami	A high-speed, low-amplitude, long-wavelength gravity wave that is triggered by a submarine seismic event or by the eruption of a marine volcano.

Wave group	A collection of waves of similar wavelengths that interfere to produce a succession of enhanced peaks.
Wave height	The vertical distance between a wave crest and the neighboring trough.
Wavelength	The distance between crests of a wave.
Wave steepness	The height of a wave divided by its wavelength.
Weak wave	A wave whose height is very much smaller than its wavelength, for which Airy theory applies.

Index

Page numbers in *italics* refer to figures, tables, and photos.